CONTENTS

About the Author	5
INTRODUCTION	6
MEGAPROJECT MANAGEMENT	11
A FUNDAMENTAL CONCEPTUALIZATION OF MEGA INFRASTRUCTURE CONSTRUCTION	17
SIGNIFICANT ENGINEERING AND CONSTRUCTION OF MEGA- INFRASTRUCTURE	23
MANAGEMENT OF MASSIVE INFRASTRUCTURE CONSTRUCTION	32
MANAGEMENT OF CONSTRUCTION PROJECTS TO MEGA INFRASTRUCTURE CONSTRUCTION MANAGEMENT:	48
THE DIFFICULTY OF BUILDING AND MANAGING MEGA INFRASTUCTURE	60
THE GOOD, THE BAD, AND THE BEST OF MEGAPROJECTS	72
MAKING MEGAPROJECTS MODULAR	82
8 PLANNING AND ORGANIZING BEST PRACTICES FOR MEGAPROJECT SUCCESS	95
THE CAUSES AND CURE FOR UNDERPERFORMING MEGAPROJECTS	99

THE INCREASING DIFFICULTY OF BUILDING MEGA- 102
INFRASTRUCTURE

THE POSSIBILITIES AND HOPE FOR THE FUTURE 111

MEGA PROJECT MANAGEMENT

Culture, Economy, and Society

By

Lim Guan Leng

Acknowledgement

I'm eternally grateful to my mum Madam Goh A.H. She taught me discipline, love, respect and so much more that had helped me succeed in life.

I would like to express my gratitude to all my previous bosses, especially Mr. Brian, who entrusted me to a few mega projects in Central Asia. These experiences widened my vision and allowed me to improve on my Project Management skills especially dealing with a multi-culture workforce. I learned to cope with engineering difficulties due to environmental or political factors.

I am thankful to my editor Tanha Emita for making important suggestions that enhanced this edition.

Lastly, I would like to say thanks to people who helped me along the way during the execution of the project overseas. We had overcome the difficulties and had made the projects a success and we are proud of what we had achieved.

Copyright © [2022] [Lim Guan Leng]

All rights reserved. No portion of this book may be reproduced in any form without permission from the author.
For permission, Contact: limguanleng.76@gmail.com

ABOUT THE AUTHOR

Lim Guan Leng is an engineer, construction expert, explorer, and author. He has worked with an engineering procurement and construction company and has megaproject management experience for about 15 years. Mr. Lim has traveled extensively and engaged with numerous projects, both at home and overseas, that made him a prescient megaproject expert and helped shaped the contents of this book.

When he is not involved in giant multi-million budget projects, you can find him hiking and roaming around in nature. He is fluent in five languages including Thai, English, Chinese, German, and Russian. Mr. Lim currently resides in Germany where he is using his free time in sailing his yacht and occasionally writing about old and new experiences.

INTRODUCTION

A megaproject is a large-scale investment endeavor. According to the Oxford Handbook of Megaproject Management, megaprojects are large-scale, complicated undertakings that often cost 1 billion dollars or more, take years to conceive and build, and includes several public and private players. These transformational projects influence the lives of people and affect nature. However, 1 billion is not a limiting factor in defining megaprojects. In underdeveloped countries, a parallel approach of far less expenditure on infrastructure (~100 million dollars) may earn such appellations. Thus, Megaprojects can be defined as transient initiatives (i.e., projects) characterized by a significant investment commitment, enormous complexity (particularly organizationally), and a long-lasting influence on the economy, the environment, and society.

Megaprojects account for 8% of global GDP- according to Bent Flyvbjerg, a professor at the Business School of the University of Oxford. Megaprojects include building and decommissioning projects with a high level of innovation and execution complexity. They are impacted by a variety of techno-socio-economic and organizational issues.

Planning bias

A bit of cognitive bias is the planning fallacy which affects our critical thinking and decision-making ability. The planning fallacy is discussed in greater detail in its' delegated segment. During the project development process, one must exercise guardianship to avoid optimism bias and strategic misrepresentation of the project. A curious paradox is on the rise that exhibits an increasing number of megaprojects, proposed despite their consistently poor performance against initial budget, schedule, and benefit forecasts.

Downsides

Corruption is frequently a factor in megaprojects, resulting in increased expense and decreased benefit.

According to the European Cooperation, Megaprojects in Science and Technology (COST) are defined by their "high complexity (both technological and human) and a long track record of poor performance." Megaprojects garner considerable public attention due to their significant impacts on communities, the environment, and budgets, as well as their enormous expenditures. Megaprojects can alternatively be defined as "physical, enormously expensive, and public undertakings."

Examples

Megaprojects are part of the development scheme. They are administered into special economic zones, public buildings, power plants, dams, airports, seaports, bridges, highways, tunnels, and railways. Wastewater treatment, oil, and natural gas extraction projects, aerospace projects, weapons systems, information technology systems, large-scale sporting events; all could be part of specific development segments. The most common megaprojects fall into hydroelectric facilities, nuclear power plants, and sports event categories. Megaprojects can also refer to large-scale, high-cost scientific research and infrastructural endeavors.

For example, sequencing the human genome is a significant global accomplishment in genetics and biotechnology.

According to Bent Flyvbjerg, "As a general rule, 'megaprojects' are worth billions of dollars,' Major projects' are worth hundreds of millions, and 'projects' are worth millions or tens of millions."

Rationale

Numerous common megaprojects are developed to achieve collective benefits, for example, power for everyone (who can pay), road access for everyone (who has a car), and so forth. They may also be used to expand frontiers. Megaprojects have been chastised for their top-down planning methods and unintended consequences for particular areas. Large-scale projects frequently benefit one group of people at the expense of another. For example, China's Three Gorges Dam, the world's largest hydropower project, forced the relocation of 1.2 million farmers. In the 1970s, highway revolts in several Western nations saw urban activists oppose government plans to demolish buildings to make way for motorway routes, claiming that such demolitions would unfairly penalize urban working-class residents while benefiting commuters. Anti-nuclear rallies in the United States and Germany against proposed nuclear power stations stalled development owing to environmental and social concerns.

Recently, new megaprojects have been identified that are no longer monolithic and solitary in goal but rather are highly adaptable and diverse, such as waterfront redevelopment proposals that appear to provide something for everyone. But some similarities with the traditional ones remain. The new ones preclude "A wide array of social practices, perpetuating rather than relieving urban inequality and disenfranchisement." Megaprojects stifle the establishment of oppositional and contestation behaviors due to diversity in land utilization. Projects are reduced to collective benefits from an individualized type of public benefits, which are frequently at the heart of a mega-reasoning.

According to Flyvbjerg, policymakers are drawn to megaprojects for four reasons:

- Technological sublime: The euphoria that engineers and technologists experience when they work on enormous, inventive projects that push the frontiers of what technology is capable of.
- Political sublime: The euphoria politicians experience from erecting monuments to themselves and their causes.
- Sublime economics: The joy experienced by business people and labor unions due to the profits and jobs created by megaprojects.
- Sublime aesthetics: The joy designers and those who value good design derive from the building, using, and looking at something enormously beautiful.

Economics

Proponents of infrastructure-based development argue that large-scale projects should be funded to generate long-term economic advantages. Since the 1930s economic crisis, investing in megaprojects to stimulate the general economy has been a popular governmental approach. Recent examples include China's 2008–2009 economic stimulus program, the European Union's 2008 stimulus package, and the 2009 American Recovery and Reinvestment Act. Megaprojects frequently raise finance based on anticipated returns – while projects frequently exceed budget and timeline expectations, market factors such as commodity prices may shift. During the planning phase, critics of megaprojects sometimes voice concerns about cost overruns. It happens due to a lack of accountability and risk-sharing systems. Bent Flyvbjerg has emphasized the existence of incentives to overstate income, underestimate costs, and inflate future social and economic advantages. When the megaproject is carried out in a country high of corruption, the risk and number of overruns increase.

Obtaining sufficient money is one of the most difficult components of megaprojects. Alan Altshuler and David Luberoff discovered that innovative and politically astute leadership is required to acquire resources while generating public support, appeasing critics, and resolving disputes throughout many years of planning, authorization, and execution. Additionally, people designing megaprojects confront problems such as empowering community organizations, challenging data and methodology, high degrees of uncertainty, minimizing negative effects on neighborhoods and the environment, and attempts to solve a wicked problem.

MEGAPROJECT MANAGEMENT

Megaproject management is a relatively recent addition to the discipline of project management. Back in time, it was initially considered a subset of project management. A continuing chain of high complexity and history of associated failure has picturized an urgency to evolve it into a distinct specialization. Bent Flyvbjerg, previously recognized- Developing the first applied process appears to be the main factor in classifying it as a distinct field. Megaprojects prompted this procedure.

History
Mega is meaning for large. Imagine a mega-city rather than a mega-joule when considering a megaproject. It initially corresponded to the unit prefix, as such projects were predominantly dams, bridges, or very massive structures in the early twentieth century.
The next upward trend occurred with the Manhattan Project and then with the Apollo program- both are cited widely as positive examples. The term 'Megaproject' gained popularity in the 1970s, as project costs surpassed the billion-dollar mark. Currently, projects costing between $50 billion and $100 billion are popular. We have reached the trillions benchmark considering certain activities that require devoted administration. Such as the 2008 stimulus packages or US defense procurement. We are approaching a

'Tera era' of megaprojects.
Leaving these exceptional occurrences aside, Megaproject's account for 8% of global GDP when infrastructure and industries with billion-dollar projects are included.

The Four Sublime
These are the four factors that contribute to megaprojects' appeal. They are a form of group prejudice for each stakeholder category. They are as follows:
- Technological sublime: Engineers and technologists constantly strive to create the latest/tallest/fastest things.
- Political sublime: Because politicians adore the opportunity to be associated with large-scale projects and the publicity that entails.
- Sublime economics: Because unions, contractors, and businesspeople adore the jobs and money.
- Sublime aesthetics: Designers enjoy creating beautiful things, and the audience enjoys adopting large beautiful items as symbols of their city/country.

As is natural with biases, there are consequences:
When the four sublimes are at work, and the megaproject format is chosen to accomplish large-scale initiatives, the following qualities of megaprojects are frequently disregarded or glossed over:
1. Due to their lengthy planning horizons and intricate interactions, megaprojects are inherently dangerous.
2. Frequently, projects are led by planners and managers with limited domain expertise who may change during a project cycle, leaving the leadership vulnerable.
3. Typically, decision-making, planning, and management are multi-actor processes involving a diverse range of public and private stakeholders who have competing interests.

4. Because technology and designs are frequently non-standard, planners and managers develop a "uniqueness bias" in which they view their projects as unique, obstructing their ability to learn from other projects.
5. Early commitment to a particular project concept is often excessive, resulting in "lock-in" or "capture," obviating the need for alternate analyses and escalating commitment in subsequent stages. "Fail Fast" is not applicable; "Fail Slowly" is (Cantarelli et al., 2010; Ross and Staw, 1993; Drummond, 1998).
6. Due to the enormous sums of money involved, principal-agent conflicts and rent-seeking conduct, as well as optimism bias, are prevalent (Eisenhardt, 1989; Stiglitz, 1989; Flyvbjerg et al., 2009).
7. The project scope or ambition level will often alter dramatically over time.
8. Delivery is a high-risk, stochastic activity that exposes individuals to so-called "black swans," or severe events with catastrophic consequences (Taleb, 2010). Managers frequently overlook this, approaching projects as though they were largely predictable Newtonian universes of cause, effect, and control.
9. Statistical research indicates that such complexity and unforeseen events are frequently overlooked, resulting in insufficient budget and time contingencies.

As a result, misinformation about prices, timetables, benefits, and risks pervades the project development and decision-making processes. As a result of cost overruns, delays, and benefit deficiencies, the project's viability throughout implementation and operation is jeopardized.

Megaprojects' Iron Law
- With time.

- Excessive expenditure.
- Insufficiently utilized.

These are not insignificant costs: cost overruns of 1.5x are frequent; in extreme circumstances, they can exceed 10x, and 90 percent of projects experience them; it is also normal for projects to have a utilization rate of 0.5x or less once completed. This holds for both the public and commercial sectors and across countries. Thus, arguments such as excessive regulation or corruption are inadequate. They begin poorly but eventually succeed. The reasons are illustrated below.

Break-Fix Model
Because the administration of megaprojects has no idea what they are doing or lacks the motivation to care, something will go wrong. Then the additional time and money are spent repairing what went wrong, or the project's terms are renegotiated, and the project limps onward to the next setback. This method is repeated until the project is completed.
If it is so heinous, and we know that it is heinous, why do we do it this way?

Hirschman's Hiding Hand
Because many critical stakeholders are unaware of how bad it is. From Willie Brown, former San Francisco mayor:
"The news that the Transbay Terminal is approximately $300 million over budget should come as no surprise. We were always aware that the initial estimate was significantly lower than the actual cost. Similarly, we never received an accurate cost estimate for the [San Francisco] Central Subway, the [San Francisco-Oakland] Bay Bridge, or any other large-scale building project. Therefore, get off it.

In the field of public works, the initial budget is essentially a down payment. Nothing would ever be accepted if people were aware of the true cost from the start. The objective is to get started. Begin by excavating a hole that is so large that there is no other way to fill it in except to come up with the money to do so."

Nor are they without reason, as arguments in support have been given. The first argument is identical to Willie's: *If we realized how difficult large undertakings were to complete, we would never undertake them.*

For the second, note that the section's title is obscured, not obscured. Albert O. Hirschman made this argument based on earlier work by J.E. Sawyer, and it states that there is an error in both cost and benefit calculations and that this error should nearly cancel out. The issue is that Sawyer's study just demonstrated that this was conceivable based on a randomly chosen sample of five or so. Hirschman subsequently expanded it into a "Law of the Hiding Hand," legitimizing self-deception.

Regrettably, that is untrue. Apart from being refuted by actual statistics, Flyvbjerg illustrates the non-monetary opportunity costs of the Sydney Opera House. Dane Jorn Utzon, the architect, won the Pritzker Prize (architecture's Nobel) in 2003 for the Sydney Opera House. His lone significant work is the project's disastrous delays. Cost overruns effectively ended his career. In comparison to another gifted architect Frank Gehry, it appears as though management's mismanagement of the Opera House cost us half a dozen magnificent landmarks.

Survival of the Inadequate

The prevalent mindset that it is completely okay to fabricate and then mismanage megaprojects creates an odd situation in which fewer desirable projects are more likely to be selected. Consider two competing projects, one led by an honest and competent manager and an incompetent and dishonest manager. The latter project appears to have lower costs and greater advantages. Those deciding between them probably expect both to be over budget and behind schedule by roughly the same amount. As a result, we consistently make poor decisions regarding initiatives to build.

Light at the End of the Tunnel

Fortunately, there are some silver linings. These shortcomings were identified as a critical policy area for the US government under the Obama administration. It is increasingly much more typical for megaproject failures to result in leadership consequences, such as the CEOs of BP and Deepwater Horizon, or Airbus and the A380 superjumbo. Some megaprojects succeed and serve as models for doing it well, such as Bilbao's Guggenheim Museum. Finally, there is sporadic adoption of various excellent practices, such as reference class and independent forecasting.

A FUNDAMENTAL CONCEPTUALIZATION OF MEGA INFRASTRUCTURE CONSTRUCTION

A concept is extensive ideation or a concise summary of anything or specific operations. It takes shape in the individual's mind upon the constant endeavor of exploring a certain issue. A concept might be an abstraction of the essence and qualities of the subject under study, or it can be a delimitation of the scope of discussion for a research issue.

Clear and precise notions are critical for theoretical studies, particularly when defining those essential themes. To avoid arbitrary or unclear interpretations of critical concepts, it is compulsory to define the components precisely that are employed in the study.

This book's most critical and essential notion is of Massive Infrastructure Construction. Broad agreement and consensus on this concept and its derivative discussion *Mega Infrastructure Construction Management* have been achieved through years of construction practice. The author provides basic and essential definitions of the fundamental concepts in this book, including *Mega Infrastructure Construction* and *Mega Infrastructure Construction Management*, with the dispense of academic rigor in theoretical studies.

Engineering and Project

Engineering

Men have engaged in various material production activities dating back to the dawn of ancient civilization, including hunting, fishing, chicken keeping, crop cultivation, and fruit collection. Additionally, men launched a variety of entity-creation and entity-use activities. For instance, people have constructed buildings, roads, dams, and bridges to meet the most fundamental demands for survival. Historically, most objects employed in industrial activities were already present in the natural world. Nevertheless, an entity-forming process deliberately generates artificial entities or remolds existing organic entities to fulfill human demands better. As a result of this remolding, nature is reshaped to some extent.

In this respect, the entity-creating activities inherent in man's practical actions might be viewed as the antecedents of modern engineering activities.

Man's entity-creating activities, including those that alter the original characteristics of existing entities, are comprehensive processes that include entity-generating ideas, formulating design plans, organizing and implementing development activities, and supervising the completion of the artificial entities intended to be created. This entire process indicates that engineering is primarily concerned with constructing artificial objects or entities in response to specific human intents (Sheng and You 2007; Sheng et al. 2008).

At times, the term "engineering" refers to the *Creation of artificial things or entities*. It is simply a method of expression in Chinese that is habitual. For instance, in the line

"China's Great Wall is a magnificent engineering feat," The term "engineering project" is used to describe the physical thing known as "the Great Wall." With the advancement of man's entity-creating activities and the progressive abstraction of man's cognition and perception of the universe, the concept of engineering has taken two distinct trajectories.

Because man's entity-creating engineering efforts began with constructing houses, roads, and dams, early entity creation was characterized by characteristics of civil engineering. In light of civil engineering's purposes, it is vital to establish clear objectives for entity-creating engineering activities and to see them through to completion. The evolution and expansion of man's practical actions have broadened the field of engineering activity. It is widely agreed that human actions that incorporate comprehensive engineering procedures and have specific objectives, can be considered engineering activities in a given field. Thus, several engineering principles, such as mechanical engineering, chemical engineering, electrical engineering, and computer engineering, have been advocated over the years. With the integration of several science fields into educational institutions, these engineering principles have steadily evolved into matching disciplinary notions in postsecondary education.

With further social and technological advancements, humans have expanded the entity-creating concept of engineering to encompass additional fields, including social science, technology, psychology, culture, education, and logic, resulting in the emergence of concepts with richer meanings, such as software engineering and systems engineering.

The definition of engineering, which originally included civil and hydraulic engineering, has changed and expanded significantly, currently covering mechanics, electronics, and information science subjects.

Additionally, the scope of engineering has been expanded from physical to semi-physical and nonphysical engineering. This progression demonstrates the ongoing advancement of man's entity-creating methods and the continuous extension of their scope. This idea's continual evolution helps people explain and comprehend those human-specific entity-creating actions and processes. Once the fundamental definition of engineering is grasped and the pertinent background information is considered, the precise definition of engineering in a particular situation can be presented simply and accurately. As stated previously, engineering, especially massive infrastructure construction, as investigated and addressed below, relates exclusively to the engineering disciplines that create physical entities, such as highways, bridges, and hydraulics.

Project
The concept of a project is inextricably linked to the concept of engineering. In general, a project is a collection of unique and intricate tasks tied to one another. These operations typically have a defined objective or purpose and necessitate the allocation of specific resources, adherence to certain standards, and adherence to well-defined time frames and budgets. As such, a project could be the construction of a house, the development of a product, or the organization of an event.
Because the objectives and purposes of various projects vary substantially, they are classified into two categories: physical and nonphysical. In terms of physical purposes,

the definition of a project is almost identical to that of physical entity-creating engineering activities, with few significant distinctions. As a result, the following is safe to state:

1. When it is unclear whether a project is physical or nonphysical, it may encompass both semi-physical and nonphysical tasks. As a result, the term "project" typically has a broader meaning than "engineering," which is more focused on entity-creating activities because, in many circumstances, the project may refer to actions that extend beyond the scope of entity-creating activities. For instance, cultural and educational programs transcend the limits of entity creation, such as the *Hope Project*, a national project launched in China to help rural education, and the *211 Project*, one of China's national projects established to promote college education.

2. The project concept puts a higher emphasis on the uniqueness of activity and how to acquire and use resources. It also ensures the distribution of personnel and assigning different duties to different people. Organizing diverse activities inside a project in an orderly fashion is also a significant considerable segment. From this vantage point, the project concept places a higher premium on on-site operations and organization of a specific activity than engineering, which places a higher premium on generalizing the characteristics of human activities that result in artificial entities at the macro and global levels.

3. Because a project emphasizes the mission, organization, and operation's specific activities, individuals frequently use the term project rather than engineering when they need to break work down into precise tasks and identify the specific operations. It is a typical and customary technique

to distinguish these two notions in the building business. Indeed, this distinction could be interpreted as routine use of terms that bear little specific meaning.

SIGNIFICANT ENGINEERING AND CONSTRUCTION OF MEGA- INFRASTRUCTURE

Engineering Of Significance

With the development of human society and the advancement of science and technology, construction activities have incorporated a variety of characteristics. Complex environmental conditions, large-scale projects, advanced technology, massive investments, lengthy construction periods, lengthy project life cycles, and a significant and continuing impact on the socio-economic environment are prominent considerable derivatives. These have become increasingly conspicuous over the last century. Such constructions inspired the creation of new human habitats or bartered substantial improvements to existing ones. It created a strong foundation for the long-term growth of human society and civilization, and gradually advanced the concept of important engineering. Fundamental characteristics such as enormous scale and massive influence contribute to the public's rational and intuitive knowledge of significant engineering. Indeed, the key to comprehending the concept of substantial engineering is to understand how the term "major" is perceived. For example, in terms of construction investment, the Federal Highway Administration of the United

States previously defined mega constructions as those exceeding one billion dollars in cost. At the same time, the Norwegian government established a cost threshold of 60 million euros for significant engineering projects. Given the contrasting economic conditions of different countries, determining a megastructure is solely based on the amount of investment that would result in significant discrepancy. It demonstrates that it is not feasible to define significant engineering as a quantifiable concept and quantify it quantitatively. As a result, when long-term construction methods are used as a guide, individuals are more likely to synthesize the fundamental characteristics of major engineering from many viewpoints and levels before establishing a descriptive definition. This intuitive approach to interpreting and comprehending significant engineering is widely acknowledged. The perception of a structure is highly dependent on the public's understanding of the structure. The public view is that a substantial engineering construction is enormous in scale and noteworthy.

According to the breadth of the functions that engineering can perform, the entity-creating engineering works are categorized as follows:

1. *Mega-engineering in the scientific and technological fields.*

This term refers to engineering structures devoted to the exploration and discovery of significant scientific rules or the achievement of significant technological breakthroughs to achieve significant scientific and technological development goals in a condensed period. These types of engineering, for example, joint gene and material microstructure studies conducted by multiple countries, or joint spacecraft studies conducted by one or more countries, will have a global impact and provide an overall impetus for social development by advancing man's understanding of natural laws, contributing to breakthroughs in critical technologies, enhancing the strategic competitiveness of specific industries, and so forth.

2. *Engineering on a massive scale for military and national defense.*

This term refers to substantial engineering projects undertaken by individual countries or coalitions of countries with the primary objective of researching and developing weaponry and military equipment to ensure defense security or bolstering national military capability. For example, the United States' National Missile Defense (NMD) system and Russia's Global Navigation Satellite System (GLONASS) come under this category.

3. *Construction of massive infrastructure.*

Infrastructure's most essential definition is fundamentality and "laying the groundwork for societal growth." The World Bank's World Development Report 1994 (1994) advocated that infrastructure be defined as "permanent structures, equipment, facilities, and the services they offer for human survival and social production."

Economic infrastructure is a subset of infrastructure that comprises urban public utilities, transportation facilities (such as highways, ports, and airports), and public works (such as dams and hydraulic facilities). The second category of infrastructure is social infrastructure, which comprises educational, cultural, and healthcare institutions. Physical infrastructure initiatives directly address and meet the human species' essential survival and everyday life demands. These infrastructure projects are typically vast in scope and extremely significant. Thus, such projects should be classified as massive infrastructure construction.

In comparison, mega infrastructure construction, the primary goal of improving people's lives and facilitating social development, is more prevalent than the other two major engineering disciplines, namely mega scientific and technological engineering and mega military and national defense engineering. For instance, China's South-North Water Diversion Project, the Three Gorges Dam Project, and the Hong Kong-Zhuhai-Macau Bridge Project are all examples of major infrastructure projects critical to the country's well-being and livelihood.

Mega Infrastructure Construction

Significant scientific and technological engineering and a major military and national defense engineering have respective requirements and operating procedures. Considerable objects are- environment, purpose, subject, decision-making, and execution process.

Additionally, because the relevant information is highly secret, the majority of such engineering would result in an information imbalance when disclosed to the public. As a result, studies on such engineering may demand unique research approaches and methods. In comparison, major infrastructure projects that directly affect people's livelihoods fully embody the fundamental characteristics of substantial engineering and thoroughly embody the norms of economics, management, and a variety of other disciplines. Because there is a wealth of relevant information available to the public and sufficient samples for exploration, conducting theoretical studies of these significant mega infrastructure constructions to uncover the underlying general rules and rational explanations for common phenomena in such projects is of great academic value and universal practical significance.

China is currently the world's largest developing country. With a large population to support and advance social and economic development, as well as strategies such as urbanization, China must strive to increase housing construction and accelerate the development of infrastructures such as highways, railroads, bridges, and communication facilities, as well as improve the environment through hydraulic engineering and environmental protection projects. This reality establishes China as a global leader in infrastructure development.

Additional proof for this claim may be found in an article published on March 24, 2015, by The Washington Post. According to this article, China consumed more cement in the last three years than the United States did over the twentieth century. Specifically, America consumed around 4.4 billion metric tons of cement throughout the last century, while China consumed approximately 6.4 billion metric tons from 2011 to 2013. (Ana 2015). In an interview with the Global Times on March 25, 2015, the China Cement Association officials confirmed this figure. They said that it might be attributable to China's present rapid expansion needs in various industries (Xing and Chen 2015).

According to a review of several projects, mega infrastructure constructions refer to those that encompass a broad range of activities and a significant amount of building and are widely acknowledged as the sort of project with the closest ties to the public. The most critical fundamental characteristics of massive infrastructure projects are as follows:

1. In general, the state (government) is the primary de-

cision-maker and financier of massive infrastructure projects. As a result, the state (government) frequently takes the lead in the process of massive infrastructure construction, deciding on major issues such as whether a project should be approved, whether it should be funded, and how the project should be carried out (Cairns 2004).

2. Mega infrastructure projects are frequently built on colossal scales. Most of the projects are carried out across wide swaths of land and on colossal scales. For instance, an inundation area, or a 632 km2 area of land prone to flooding, is the outcome of the Three Gorges Dam Project and encompasses over 20 cities and towns with a combined population of 850 thousand. Considering such as second-time resettlement, the dynamic population resettled by this program was 1.13 million persons (Shi et al., 2011). Additionally, the Three Gorges Dam Project was scheduled to be completed in three phases during a 17-year construction span at the cost of RMB 332 billion (Wang 1999). Another example is the West-East Gas Pipeline Project, which extends eastward from the Tarim gas region in Xinjiang to Shanghai, with the main pipeline spanning over 4000 kilometers and requiring a total investment of RMB 120 billion in the first phase (Wu 2004).

3. Complex environmental circumstances define mega infrastructure constructions. Mega infrastructure projects are frequently located in areas with complicated, if not hostile, environmental circumstances. For instance, the Qinghai-Tibet Railway project is a 1956-kilometer railway that runs from Xining, Qinghai province, to Lhasa, Tibet Autonomous Region. The majority of the project is located on the Tibetan Plateau, dubbed "the roof of the world" and "the world's third pole" (Sun 2005a). As part of the project, 960 kilometers of railroads will be constructed in places

with an average height of more than 4000 meters, resulting in anoxic atmospheres, extremely low temperatures, and complicated and changeable climates. Additionally, around 550 kilometers of the railroad were built on continuously unstable permafrost zones, defined by persistently low temperatures and thin frozen soil layers.

As a result, another obstacle in this railway project was the construction of tunnels at the world's highest height and across the world's longest mile of permafrost.

4. Mega infrastructure projects are defined by their profound and far-reaching effects on a region's social and economic growth. Mega infrastructure projects are typically undertaken to advance the social and economic development of a given region or improve society's living environment over the long term. As a result, the objectives of mega infrastructure construction must be to generate significant positive energy for regional social and economic development. However, it should be recognized that not all of the objectives stated for major infrastructure constructions can be met.

Indeed, actions made with insufficient planning or significant decision-making errors may have unfavorable repercussions. Given the far-reaching influence that large infrastructure projects can have on a region's social and economic development, such mistakes can result in enormous long-term and irreparable damage. Numerous such catastrophes have occurred in big infrastructure projects worldwide throughout the last few decades.

5. Mega infrastructure projects often have a long-life span. Due to various factors, including large construction scales and complicated environmental and technical circumstances, the duration of a mega infrastructure project, from

original design to completion phase, is frequently many years or even decades. Additionally, after a completed structure is put into service, it can last decades or even centuries until it reaches the end of its life cycle. Due to the likelihood of such long-life cycles, a mega infrastructure project may be divided into several components completed in phases. More importantly, a long-life cycle implies that numerous uncertainties in the social and economic environments, as well as in changing natural environments, will occur during the implementation of a project, particularly during the process of realizing its original goals, necessitating the functionality of a mega infrastructure project to be robust and stable over an extended period (Priemus et al. 2008). These constraints impose significant problems in terms of the quality of major infrastructure project construction and the quality of decision-making related to such projects.

6. Numerous partnerships are tasked with implementing massive infrastructure construction projects. Different partnerships may have divergent project objectives and pursue divergent interests in a large infrastructure development project, resulting in conflict and competition among the various partners during the decision-making process. For instance, vigorous public participation may result in intricate relationships between the various parties involved in constructing a large infrastructure project.

While not exhaustive of all key qualities of mega infrastructure builds, these characteristics encapsulate the popular understanding of the usual characteristics of mega infrastructure development and their judgments of its scientific implications. As a result, these characteristics can be viewed collectively as the descriptive definition of mega infrastructure construction. As covered in the following chapters and as the primary focus of this book's study, significant engineering refers to the building of mega infrastructure. The characteristics outlined earlier indicate that mega infrastructure builds possess key characteristics that are absent (or, if present, are neither unique nor noticeable) in other types of projects. Indeed, these characteristics have combined to create a new entity-creating activity pattern known as mega infrastructure construction practices, which has fundamentally enlarged and modified man's knowledge of engineering. Furthermore, it is commonly accepted that mega infrastructure construction's fundamental characteristics and properties have contributed to the establishment of a novel sort of cognitive rule. This rule compels people to recognize that mega infrastructure construction is a distinct type of engineering or engineering activity process from other engineering activities; additionally, it will profoundly affect future mega infrastructure construction management activities and theoretical research on mega infrastructure construction management.

MANAGEMENT OF MASSIVE INFRASTRUCTURE CONSTRUCTION

Construction management activities are crucial and integral components of entity-creating activity. Not only do great projects require building, but also management. Like mega entity-creating construction activities, mega infrastructure construction management activities encounter a variety of new occurrences, issues, and laws, all of which require individuals to expand their thinking and acquire new cognitions.

Overview of Construction Management

People may carry out modest, small-scale construction tasks in ancient times. Typically, a single individual could only perform the work required for a relatively straightforward entity-creating building operation. However, as building activities grew in scope, it became increasingly difficult for a single individual to accomplish the work required to complete the entire construction activity autonomously. People tended to collaborate and establish specific groups to complete specific components of the activity in such conditions. Thus, humans overcame several obstacles and finished sophisticated projects through their collective strength and intelligence. As a result, collaboration became an increasingly common trend (Cicmil et al. 2006).

Humans realized through these practical actions that

in collective entity-creating construction activities, they needed to improve their organization, distribute work, and divide the entity-creating construction process into distinct stages that were progressively connected. The separation of work improved the likelihood of achieving the construction objectives and completing the construction operations more efficiently and sequentially. Additionally, one or more individuals would disengage from the specific and direct entity-creating construction work and instead engage in activities to improve the organization and efficiency of entity-creating construction activities (Capka 2004). These activities can be summarized as follows:

When diverse groups of people collaborate on single building activity, one or more individuals specialize in the planning, procuring, and distributing resources necessary for entity development. They do this by allocating and organizing assignments for various working groups and teams and coordinating relationships among the various groups and assignments to improve the organization and efficiency of entity-creating practices. These are referred to as construction management activities; "construction management" is an abbreviation for construction management activities.

The following are the implications of construction management:

1. As a result of entity-creating construction activities, man's construction management activities affect entity-creating construction activities. In turn, these two types of activity are inextricably linked. While individuals produce structures through entity-creating activities, construction management is responsible for integrating, adapting, coordinating, and regulating the actions and connections of

people, people and things, and things. As a result, humans cannot exist without entity-creating structures, and entity-creating structures cannot be implemented without construction management.

2. Construction management is a collection of comprehensive practical actions that occur throughout the process of entity creation. Like other management activities, construction management activities are defined by a comprehensive process that includes fundamental elements such as management objectives, subject organizations, challenges, and surroundings. A complex management activity can be carried out at multiple levels and in several largely self-contained portions (Wang et al., 2014).

3. Typically, construction management operations are carried out concurrently with relevant construction activities. Once a construction activity is initiated, the associated construction management operations are initiated as well. Additionally, the associated construction management operations are likewise completed once the construction activity is completed. However, as man's construction activities become more complex, relevant construction management activities must generate early construction management ideas, make early-stage decisions, plan the project, present the arguments, and carry out operation management and post-audits for the construction management activities to fulfill and progress or regress appropriately, thus extending both ends of the life cycle of a car.

4. Each entity-creating structure is distinct and distinctive, and no two constructions are identical. As a result, many construction management activities have their distinct characteristics, which means that no single construction management model can address all problems or apply to every sort of entity-creating construction activity. On the

contrary, each construction management activity evolves through time and is carried out in unique ways for unique causes by unique people in unique locations.

5. Despite the high degree of resemblance between two structures, the detailed construction management needs are inextricably distinct. This is not just because each structure has distinct and unique characteristics. In essence, construction management is a type of activity in which some people, referred to as subjects, act on other people, referred to as objects, and interaction that demonstrates that the objectives of construction management are integrated not only with people's value judgments and value orientations, but also with their cultural dispositions, including intellectual factors such as personal preferences, behaviors, emotions, and personal habits. In other words, construction management is fundamentally about people-oriented or people-centered activities. As a result, it is unwise to place a greater emphasis on the thing than on the person in construction management activities or value the thing while ignoring the person.

6. Construction management encompasses a wide range of topics. As a result, appropriate technology, methodologies, and approaches must fulfill numerous managerial activities. Construction management, in this sense, should be both realistic and actionable to ensure that it is both effective and cost-effective. When there are several models to pick from, the most appropriate and cost-effective ones should be chosen over the unnecessary and lavish ones (Qi and Liu 2012).

7. As human society evolves; people get more proficient at constructing entities. As a result, the implications of construction management evolve and become deeper, painting an enthusiastic picture of new managerial concepts, in-

formation, and approaches "sprouting out like mushrooms after rain."

8. People have long conducted studies on construction management and built relevant education and disciplinary research systems due to the rich construction management practices. Construction management ideas, techniques, and applications have never been more numerous and fruitful than they are today (Baccarini 1996).

Mega Infrastructure Construction
Management: Overview

As previously said, because mega infrastructure building possesses basic qualities that are unique to it, management activities associated with it will undoubtedly confront several new conditions distinct from those associated with construction management activities.

As a result, these essential requirements are examined one by one in light of the fundamental elements affecting construction management activities, allowing us to better understand the critical distinctions between mega infrastructure construction management and construction management.

The initial aspect to consider is the managerial environment. Because mega infrastructure projects typically span large geographical areas and vast swaths of land, the social, economic, and natural environments surrounding them will not only change dynamically throughout the project's long life but will also encounter a variety of complex phenomena such as evolution and abrupt changes (Stergiopoulos et al. 2016). As a result, these considerations may have a considerable impact on the function design and construction of a large infrastructure project and the construction quality and function stability throughout the lengthy post-construction period. For instance, the construction process may be halted due to uncertain political and social contexts; unforeseen economic instability may result in fund chain failures, and significant changes in the natural environment may impair a structure's ability to serve the purposes for which it was planned (Salet et al. 2013). Thus, intricate management environments exacerbate the difficulty of producing high-quality structures and developing strong construction functions in mega infrastructure construction management.

The second aspect to consider is the management organization. A management body for mega-infrastructure construction comprises stakeholders who have decision-making authority, property rights, construction rights, supervisory rights, and the right to discourse. Governments, project owners, designers, contractors, suppliers, supervisors, scientific researchers, and the general public are all examples of stakeholders. This group, which would comprise several mega infrastructure development management themes, would be enormous in scope and would include a diverse set of people with varying values. Nonetheless, the participants share a common objective: successfully construct and operate a huge infrastructure project. To do this, members will work to fulfill their assigned duties during the various phases of the construction management process. At the same time, divergent values among group members may result in divergent interest demands and disputes about group activity (Shi et al., 2015). This circumstance would make it more difficult for the management body to reach consensus and develop a shared purpose and result in conflicts and competition over matters concerning their interests. Such situations necessitate strong leadership and coordination capabilities on the side of the mega infrastructure construction management body and appropriate management models and procedures.

Additionally, it is critical to guarantee that the management subjects' actions comply with established criteria and prevent the body from acting inconsistently (Miller et al., 2000). For instance, on the site of a mega infrastructure construction project, a lack of coordination may result in conflicts between the two parties regarding their respective interests and responsibilities. It may occur because coordination between the design and construction parties directly impacts on-site technical interactions. Additionally, these disagreements can be exceedingly stressful and difficult to defuse in some instances.

When confronted with complex mega infrastructure construction management environments and complex problems, management subjects frequently lack adequate knowledge, experience, and competence, necessitating the management subjects' self-directed learning to enhance their management capabilities. From another angle, a lack of knowledge and skills complicates the coordination of the actions of the many management subjects, as self-directed learning activities cannot be as simply interpreted as the behaviors of specific project owners or managers. Rather than that, self-directed learning refers to developing managerial capacities through restructuring an existing set of subjects or creating a new one. As a result, it will complicate subjects' behavioral and organizational models dealing with mega-infrastructure building management.

Thirdly, there is the management issue to consider. In comparison to construction management, mega infrastructure construction management involves a greater number of management concerns that are also more difficult. While these complex difficulties may account for a small percentage of total management issues, they require managers to invest significant time and energy. However, and perhaps more crucially, if one of the difficulties is not resolved easily and promptly, the overall development and operation of the huge infrastructure project may be severely harmed.

Three interpretations are possible for the issue above.
1. Generally, such management difficulties require expertise in various areas and fields. For instance, the location of a massive transportation infrastructure project should consider the promise of alleviating traffic conditions and the social and economic impacts on the surrounding neighborhood. Additionally, the potential consequences of natural environments, such as geological and hydrological settings, on the engineering construction must be examined and if the construction would cause harm to the natural environment. Under such situations, it is unquestionably vital for specialists from many fields to pool their expertise to address such management.
2. Typically, the lines between managerial difficulties are blurred. Apart from obvious input/output relationships between various factors relating to a particular issue, association relationships cannot be determined with certainty. These include overt and identifiable association relationships and covert and unidentifiable association relations. Additionally, some of the relationships or significant aspects found and confirmed by us may change due to the influence of other factors throughout the real functioning process, leaving people perplexed and doubtful about the pertinent issues.
3. It is frequently challenging to articulate management concerns following a concise and explicit structured

manner (model). Indeed, managing large infrastructure projects frequently requires consideration of social and economic aspects, engineering and technology elements, and human behaviors and cultural values. Engineering technology factors are dominated by natural sciences and technological principles, enabling them to be described using structured models. However, in most circumstances, unstructured models are required to capture aspects such as human actions and cultural values (Liang and Sheng 2015). From this vantage point, combining structured, semi-structured, and even unstructured models is required to express these types of management issues simply and precisely. This integrated approach increases the complexity of modeling these management concerns and significantly complicates the integration of many models.

A good example of this is China's South-North Water Transfer Project. This massive inter-basin water transfer project is of strategic importance and designed to divert water from the Yangtze River's upper, middle, and lower reaches to the northeast, the Huaihai Plain area, and the northwest regions that are experiencing severe water shortages. The water is channeled through three routes, eastern, central, and western by the Chinese territory's physical peculiarities. The water transfer channels for this project traverse seven middle and western provinces from south to north, covering a total distance of more than 1000 kilometers. Regarding the relationships between the critical construction aspects, what needs to be done is to develop transfer waterways and divert clean water directly from the south to the north. The following investigations demonstrate, the reality is considerably more difficult than that:

1. Water, particularly pure water, is an exceedingly valuable and limited resource in China. If a culture of water conservation is not promoted across society, and if situations such as excessive water use and major water waste persist unabated, the project's benefits will be negated. Thus, facilitating situations necessitates a paradigm shift in industrial and agricultural output and people's patterns of living. Without modification, the limited amount of clean water available for transmission will ultimately serve no productive purpose.
2. Currently, the water sources that supply the southern region's water supply are severely polluted. For example, if pollution worsens in the Yangtze River's nearshore sections in the east and central regions, the water supply for local cities faces grave concerns of contamination. Along with pollution prevention measures, it is vital to maintain clean water sources and to guard against the northward spread of biological diseases induced by pollution. On the other hand, the project's western water supply regions are located in the higher reaches of the Yangtze River, where grassland degradation and desertification are severe. Thus, water transfer has no value for water conservation and, worse, may damage the natural refuge zones in the Yangtze River's upper reaches.
3. The South-North Water Transfer Project is a 1,000-kilometer passageway. The regions adjacent to the transfer routes and the water-receiving regions face severe water pollution, and ecological environmental degradation necessitates immediate action.
4. In addition to addressing water pollution directly in water source regions, a long-term sustainable adjustment to the economic structure is required, limiting the migra-

tion of high-polluting and high-water-consuming enterprises to this area to avoid adding to the already fragile ecological environment in water supply regions.

The construction management of the South-North Water Transfer Project entails not only the construction of water transfer channels and the establishment of direct input/output relationships between water supply areas and water receiving areas, but also the establishment of complicated associations and relationships among society, economy, and water ecosystem, all of which have an excessive amount of causality and relevance. These links and relationships extend much beyond the purely technical area of hydraulic engineering. Indeed, they have evolved and resulted in many challenges ranging from environmental governance to social and economic growth, industrial restructuring, regional development strategies, and even the modification of people's living habits and cultural ideologies. All of these issues will certainly have a significant impact on the South-North Water Transfer Project's target design, scheme design, interest coordination across numerous stakeholders, and project benefit evaluations.
Thus, for the South-North Water Transfer Project to operate efficiently over the long term, it is critical to creating improved management concepts and plans. For instance, could the water-receiving areas hiccup the water-supply areas' feeding? Only when the water-receiving areas can sustain relatively healthy ecological settings following the transition of development patterns and living modes can the South-North Water Transfer Project's initial aim of benefiting both the country and its people be accomplished in the long run.

The fourth aspect is organizational structure. For routine

constructions, where management problems are relatively straightforward and the management body is sufficiently competent, it is usually sufficient to plan and establish a management organization and delegate the responsibility of addressing all management issues that arise throughout the construction management process from start to finish. The management body may face substantially more intricate issues when it comes to managing major infrastructure development projects. It is frequently the case that the management body is incapable or insufficiently equipped to address these issues. As a result, it is challenging to develop a single construction management organization capable of analyzing and resolving all management issues that arise during the construction management process. As a result, a management organization for mega infrastructure construction should have some flexibility and adapt to potential changes in the management process, such as changing the group's composition and altering the management mechanism and procedures, to improve the overall management capacity (Ponzini 2011). For instance, the original connection between the owner and research and development institutes is typically one of direct principal-agent. However, in succeeding phases, the principal-agent connection is established indirectly via the designers and contractors. Similarly, owners and subcontractors may initially contact via contractors but eventually form direct professional subcontracting relationships, and so forth.

Fifthly, there is the managerial purpose to consider. Solving management difficulties is always geared toward management objectives. Due to mega infrastructure's lengthy development life cycle, management objectives may vary according to time scales. Additionally, because large infrastructure projects have a considerable impact on the social and economic contexts, management objectives vary by sector. Thus, it is not difficult to see how these objectives with varying dimensions and objectives with the same dimensions but different scales have not only resulted in multilayered, multi-dimensional, and multi-scale objective systems, but have also created difficulties when describing the objectives, such as vagueness, uncertainty, conflict, and immeasurability, multiplying the difficulties associated with conducting a comprehensive analysis and evaluation of the objectives (Gao and Liu 2005; Md. Masrom et al. 2015). For example, the fundamental goals of project bidding and procurement are to purchase qualified contractors, high-quality building supplies, and acceptable technologies. As a result, numerous criteria must be considered during the bidding and procurement procedures, including price, quality, and business reputation. However, trustworthy technology is most urgently required for mega infrastructure construction. Without such technology, massive infrastructure construction objectives such as quality, safety, and cost-effectiveness cannot be met. As a result, the focus must be given to modern and trustworthy technology during the actual bidding and procurement processes for massive infrastructure undertakings. This is in stark contrast to other types of construction bidding evaluations, in which business points account for a bigger share of the total. This technology-first approach should be well-implemented in the bidder assessment system for major infrastructure building. Additionally, it is not uncommon for various management objectives associated with a particular mega infrastructure construction to conflict with one another and for some of those objectives to be difficult to quantify.

Sixthly, there is the management program to consider.

The management program refers to the solutions, plans, approaches, and methodologies offered to solve the many management challenges associated with massive infrastructure construction. For these relatively straightforward issues in mega infrastructure construction management, developing a management program is similar to that of other types of construction. However, the development approach of a management program for really complicated problems is somewhat different.

To begin, the management subjects' understanding of a complex problem undergoes a certain process as a result of human cognitive rules, namely, from knowing nothing to knowing a little, from knowing a little to knowing some, from knowing some to knowing superficially, and from knowing superficially to knowing profoundly. This process reflects the depth of the subjects' understanding of the problem and enables the management body to reach a consensus as a group (Guo et al., 2012).

Thus, while they collaborate to establish a management program for mega infrastructure constructions, the managing bodies undergo an exploratory trial-and-error process. Management programs are often not produced instantly and are not presented as optimal during this phase. Rather than that, a management program is typically verified after several comparisons to and revisions and enhancements to several alternative management programs, all of which are based on the management group's depth and accuracy of understanding of the problems to be solved (Salet et al. 2013). In other words, the process of developing a management program involves incremental iterations, approximations, and convergences before the final program is produced. It begins with periodic draft solutions and progresses gradually from ambiguity to clarity, from partiality to comprehensiveness, and poor to high quality. As a result, it is unavoidable that several new and intricate links and interfaces will occur during the scheme's development. For example, increased coordination and communication between managing subjects are required; additional revisions and version comparisons are required as draft schemes iterate; various types of information are effectively integrated; and overall evaluations and optimizations of the schemes, including cost, time efficiency, and quality, are performed.

As indicated in the following pattern, throughout selecting a program, it may be recognized that the initial program is undesirable in terms of overall technological and economic impacts or that the design objectives vary from the original project objectives. In such circumstances, the scheme study should revert to the previous phase and create a new scheme for in-depth examination to identify a final program (Fig. A).

To summarize, regardless of which aspect of management activities is analyzed, we observe clear distinctions between mega infrastructure construction management activities and other construction management activities and the emergence of numerous new intuitive features. Additionally, these characteristics cannot be viewed as a quantitative accumulation of the same characteristics of construction management activities, such as larger activity scales and a greater number of subjects. Indeed, they demonstrate that new qualities emerge in management activities, such as the adaptability and flexibility of management organizations and the iterative processes associated with developing management schemes. The data demonstrate that these new features indicate substantial shifts in the scientific implications of construction management from small-scale infrastructure building to large-scale infrastructure construction. Simultaneously, these developments compel us to perform new and in-depth analyses of mega infrastructure building management techniques and theories, in addition to inducing a series of management reforms.

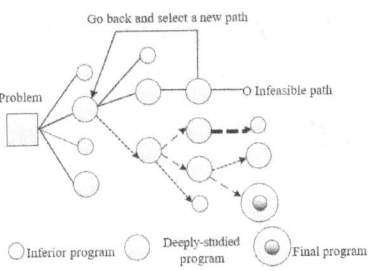

Fig. A. Program comparisons based on feedback and supplements

MANAGEMENT OF CONSTRUCTION PROJECTS TO MEGA INFRASTRUCTURE CONSTRUCTION MANAGEMENT:

From Systematisms To Complexity

After comparing the management of mega infrastructure construction and regular construction and identifying numerous intuitive distinctions between the preceding section, the next step toward comprehending the scientific connotation of construction management and mega infrastructure construction is to ascertain the two constructions' respective properties or natures.

Given that construction management is founded in construction and mega infrastructure construction is anchored in mega infrastructure construction, it is necessary to define the properties of construction and mega infrastructure construction before identifying their individual properties. The properties of an entity are the distinct qualities and characteristics that the entity reveals during its life and operation and serve as the fundamental distinction between it and other objects. Only until an individual comprehends the substance of these traits and characteristics can he or she comprehend building and mega infrastructure construction management, as well as their particular properties.

The Concepts of System and Systematisms

Since ancient times, humans have been focused on developing an overarching vision of the world that enables them to see the links between all things based on practical production, engineering, and social activities. This resulted in the emergence of fundamental worldviews such as ancient China's harmony with nature and ancient Greece's perspective on an atomic world (Qian 1981), which reflect the ancient people's total lack of comprehension of the world. This mode of cognition is likewise characterized by straightforward, organized reasoning.

Since the seventeenth century, European philosophers and scientists have proposed universe-related hypotheses. Newton, for example, believed that the universe is an interconnected system, whereas Leibniz proposed that everything occurs within a system and that its relationship to the system determines its nature. Kant, a German philosopher, was the first to establish a theory of human knowledge's systematicity (Qian 2011).

Since the nineteenth century, scientific conceptions of systems have gradually developed. The most representative of which is one emphasized by Austrian research biologist L.V. Bertalanffy (1951, 1968), who asserts that organisms should be treated as a unified natural whole that is limited in time and space and complex in structure. Bertalanffy described a system as a complicated structure comprised of several interconnected pieces.

From a broad perspective, the term "system" refers to an

abstract idea for the fundamental quality of the objective universe and human actions. That is, a system, as the most fundamental and fundamental idea, reflects and generalizes the facts and features of material items' universal connectedness and integrity. The term "system" refers to an integrated entity that performs specific functions and is composed of interconnected, interactive, and reciprocal components. In this objective world, such systems exist universally. To be more precise, a system as a whole is formed of several pieces (individuals) that interact in various ways and contribute to the system's overall structure, holistic conduct, property, and function. A system's behavior, properties, and function are typically dynamic. Additionally, a system has a diverse range of interactions and interrelationships with its environment. In contrast, system science studies the relationships between the parts and the whole and the local, global, and hierarchical relationships of things in the objective world from a system's perspective.

A system's three fundamental notions are its environment, structure, and function (Bertalanffy 1975, 1981). The term "system environment" refers to what is external to the system, whereas "system structure" refers to what is inside the system. One of the most critical qualities of a system is that its attributes and functions and its integrity extend beyond the components, with the system structure serving as the external expression. This indicates that understanding each component of a system does not imply knowledge of the system as a whole because the system's integrity is determined by the emergence of the system as a whole, not by the qualities of its components. According to system research, "it is a fundamental system principle that the structure and environment of a system, as well as their interactions, influence the system's integrity and performance" (Yu 2014).

"According to the preceding system concept, to achieve the functions we anticipate of systems, particularly the best functions, we can express alterations and adjustments to the system's structure, environment, or relationships. However, we cannot ascertain whether or not the system environment is modifiable. If it is impossible to alter it, active adaptations should be made. Nonetheless, the system's structure can be altered, adjusted, designed, and organized to achieve coordination and collaboration through the alteration and adjustment of system components or the relationships between system components and hierarchical structures, as well as between the system and its environment. The integrity of the system enables the realization of optimal functions, which is the fundamental concept of system organization and management, system control, and system intervention." (2014) (Yu).

By reiterating the preceding set of statements and definitions of a system, it is clear that a system, in a broad sense, possesses the following intrinsic and unique features (properties):

1. Diversification (every system is composed of multiple elements)
2. Correlation (the varied elements of a system are correlated)
3. Sincerity (the existence, behavior, function, etc. of a system are coherent with each other)
4. Adaptivity (the state and behavior characteristics, etc. of a system are in flux)

Combining these characteristics, namely diversity, correlation, integrity, and dynamics, is referred to as systematisms. Because the other attributes contribute to and are a component of the system's integrity, the system's integrity is regarded as the most crucial property.

Systematic thinking, and in particular integrity, has dramatically altered the way we think and solve problems, as demonstrated by the following:

1. In many circumstances, the object under study should be analyzed holistically to determine its structure and purpose and to evaluate the link between the whole, its constituent pieces, and the surroundings.
2. In addition to classic research approaches such as analysis, decomposition, and anatomy, a strong emphasis should be placed on examining correlations, the integrity of the problem, and its relationship to the external environment.

The ideas of system and systematicity and the shifts in thought that they elicit are critical for understanding construction and construction management, and they also have significant instructive value in terms of their qualities.

Construction and construction management are both systematic.

The substantive qualities (properties) of construction and its administration are investigated through the lens of system science, particularly the system definition.

Each construction entity is a holistic entity formed of correlations and synthesis of a range of material resources, such as land, capital, material, and equipment, subject to natural and technological rules. Because construction has an explicit physical hard structure that defines its fundamental physical functions and that these material resources are critical components of the overall construction's physical engineering (Williams and Samset 2010), any construction appears as a complete form of an entity system from an overall perspective, which means that any construction is a system. A construction entity system is generally considered a hard construction system that combines with the surrounding social and economic environment to generate a new construction-environment composite system upon completion of the construction.

Because creation is at the heart of construction practice. It is a continuous process of action in which hard resources are successfully integrated into the building's hard system through the formulation, design, and implementation of construction ideas. Thus, any building technique is a systematic practice.

Meanwhile, any building practice is a comprehensively organized activity comprising a range of practical components. This comprehensive activity encompasses the roles of each part of practice, the relationships between practices, the organization and interaction of practices, and the ultimate systematically integrated mode of practice (Williams 1999; Winch 2013). In other words, construction practice fully represents the system's fundamental qualities and characteristics, such that each construction practice is the system's practice.

Similarly, any construction management activity contains parts of fundamental management activities. Additionally, diverse management activities are tied to one another by particular rules and principles that reflect the management activities' function and behavior. In this sense, construction management is a system that facilitates the design and building of a physical system. We refer to the construction management system as a soft construction system to differentiate it from a hard construction system. The following key understandings of construction and construction management have been derived from the concept of the system. Every construction or construction management practice is the practice of the system, and the system is the practice of the system.

As a result, systematicity is assumed to be a substantial-quality of construction and construction management.

What does it mean for construction management to be systematic? To address this question, we must first explore the connotation of construction and the definition of the system, given that humans are naturally involved when examining the building and management of construction creation activities from a systematic perspective. This method is comprised of the following two components:

1. Examine the structure using the system. Understanding, analyzing, and resolving problems involving the management of project construction organizations in terms of the system's principle and method are substantially consistent with the concepts that construction is the activity of creation through the orderly integration of resources, that construction resources are generally the elements of a hard system of construction, and that the orderly integration of construction resources is the association and comprehensive review of construction entails treating the structure as a whole. Additionally, an overall plan, design, and structure of the practice of construction creation will be realized through the analysis of elements, correlations, functions, and organizational behaviors.

2. Conduct an engineering examination of the system. This entails adopting a modern engineering methodology to the construction of artificial systems, particularly those composed of hard resources, such as bridges, airports, dams, ports, and other mega infrastructures, because these cannot be designed and constructed solely based on experi-

ence and extensive methods, but rather based on engineering principles requiring rigorous proof and procedure. Additionally, they necessitate implementing advanced and integrated approaches for analysis, prediction, and testing to develop a scientifically sound and dependable connected artificial engineering system.

Systems engineering was coined in the mid-twentieth century to refer to the system processes of assessing, planning, organizing, and managing based on system concepts. As a result, construction management can be thought of as systems engineering construction. In general engineering, management, i.e., construction of systems engineering, is primarily concerned with planning, designing, and running the engineering by the system concept. Additionally, it includes clear objectives, rigorous analysis, an emphasis on sequencing and on a quantitative method for coordinating the engineering and the environment, and a focus on the engineering's continued maintenance of good comprehensive functions over its lifetime, as well as the gradual expansion of comprehensive functions that span engineering, society, economy, and environment, among other issues. Zhang 2009; Koontz and Weihriclh 1988).

The application of systems engineering began with hard systems of engineering. Its greatest success was realized by applying organization management to the hard system of engineering, which has a "stiff" structure due to the strong correlation between certain aspects. To a significant extent, the qualities of this form of the system can assist us in comprehending and analyzing the entire quality of engineering. Simultaneously, the structural model and optimum approach can assist us in determining the optimal engineering program.

Thus, the following summarizes the systematic nature

of construction management. Construction management is defined as the planning, design, and on-site operation of construction creation operations guided by the system concept. However, this concept of construction management is based on a clear objective and rigorous analysis. It emphasizes sequencing and a quantitative approach for achieving the construction's overall objective and complete effect. More broadly, engineering management is systematic when it adheres to the unity of integrity and correlation and considers the dynamics of activities about the engineering creation process.

The understandings mentioned above of construction and management based on system properties are critical because they advance cognition from intuitive and perceptual viewpoints to a theoretical perspective on their substantive properties. As such, an epistemology based on system abstraction can be established beyond the concreteness of construction and its management, and thus a more detailed description of the fundamental definition and relative theories of construction and construction management can be provided through the application of logic and discourse systems from the scientific thinking system. This will establish a conceptual framework and criteria for an improved understanding of mega infrastructure building and management.

The Concepts of Complex Systems and Complexity
As human society grows and its degree of organization increases, the system's key elements get more numerous, the types of linkages between these elements become more complicated, and the formation route and change patterns of the system's overall performance become increasingly diverse. Thus, based on the initial concept, humans de-

velop a perceptual intuition for a complex system, i.e., a complex system. However, the term "complex system" may imply that the system is formed of numerous elements with varying attributes and that the connections between those elements may fluctuate. This implies that a system's complexity is determined by its noumenon or phenomena. Another possible explanation for such complexity is that the cognitive subject does not fully comprehend and understand the internal structure and correlations of the system; that is, the subject either has a cognitive ability deficit or the cognitive ability deficit is caused by the system's ontology and subjective cognition (Anderson 1972; Ladyman et al. 2013). Naturally, asserting that a system is complicated, as we frequently do, begins with a subjective perceptual cognition. For example, being unsure of the system's fundamental components, not understanding how these components are related, and being unable to appreciate the system's behaviors, phenomena completely, and developing trends that contribute and construct intuition and feelings.

The scientific idea of a complex system has gradually developed in the field of system science due to examining and summarizing the characteristics of a large number of complex systems. This concept in system science is produced through the abstraction of one type of system's attributes. Though the two expressions appear to be the same, they are not. At the moment, the fundamental definition of the term "complex system" is as follows:

For one type of system, if its components exhibit heterogeneity and self-adaptability, it is hierarchical. Its overall behavior and function cannot be generated simply by adding the components' behaviors and functions; the system is complex.

Thus, a complicated system is a subcategory of a system that is unique, and this attribute of a complex system is referred to as complexity.

Now, how do we represent complexity?

To begin, complexity is a scientific term derived from the ordinary phrase "complex." It refers to the quality of complex systems. Complexity is defined separately at the epistemic and ontological levels. Although complexity is generally associated with subjective cognition and objective connotation, it is viewed here in terms of its ontological qualities (Chen and Liu 2014; Simon 1991).

According to their research questions and the phenomena of various disciplines, such as physics, chemistry, and biology, various descriptions and definitions of the scientific term complexity have been provided, resulting in diversified definitions of the term and reflecting the natural phenomenon of varied discipline-specific connotations.

After abstracting the fundamental properties of a large number of complex systems and, in particular, taking into account the requirements of mega construction management research, we summarize the major sources and manifestations of complexity, with a particular emphasis on the context of mega construction:

1. The system and its environment exhibit a high degree of openness and interaction. As a dynamic, uncertain, and developing system, the environment is inextricably linked and interactive.

2. A system is composed of many units that exhibit variability and flexibility. This implies substantial differences between the aspects of an element's properties, effects, and functions. These aspects can actively adjust their states and behaviors in response to received information to adapt to changes in the environment and maximize their chances

of survival and development. Simultaneously, new rules will be established, and new circumstances will be formed within the system, resulting in the system exhibiting more advanced and ordered overall behaviors and functions.

3. In general, systems exhibit complicated overall behaviors and functions do not present in individual system pieces. Such behaviors and functions cannot be created by merely combining the behaviors and functions of individual system elements but must be achieved at a higher or system-wide level. This phenomenon is referred to as system behavior and function emergence. Emergence is a critical aspect of complexity (Rocha 1999; Reuven and Shlomo 2010).

Complexity has a profound effect on how we see and solve challenges. These alterations present themselves as follows:

1. A system's complexity is determined by the complexity of its constituent pieces and their linked structure. Thus, it is critical to decompose and comprehend system features from top to bottom, comprehend the partial structure, and analyze and regulate mega and micro behaviors in aggregate.

2. The complexity of complex systems is frequently the outcome of the formation of complex behaviors and relationships between system constituents. As a result, mega and micro components should be combined to demonstrate a complete grasp of the macrosystem's entire complexity. (1) is considered the reduction strategy, whereas (2) represents the comprehensive approach. Additionally, a combination of the two strategies should be used when solving problems involving complex systems.

THE DIFFICULTY OF BUILDING AND MANAGING MEGA INFRASTUCTURE

When attempting to comprehend the fundamental aspects of mega infrastructure creation and administration, a complex system and the concept of complexity are critical.

The Difficulty of Building Massive Infrastructure

Sect. 1.2.2 is a summary of the fundamental characteristics of mega infrastructure construction. If a further application of system analysis is undertaken to examine mega infrastructure construction systematically, it will disclose the following characteristics:
- The construction environment for large-scale infrastructure is dynamic and open.
- The primary structure of mega infrastructure is multifaceted, heterogeneous, generally self-contained, and adaptive.
- The critical components of mega infrastructure construction are inextricably linked, and their interactions take on various complicated forms.
- The process of constructing mega infrastructure is a synergy of system organization and self-organization.

By comparing the characteristics of mega infrastructure building to the fundamental notion of a complex system, it is possible to conclude that mega infrastructure development is, in fact, a complex system or complex construction system. In other words, massive infrastructure construction can be safely considered a subset of an artificial complex system, with the physical construction entity serving as the complex system's hard system.

Because mega infrastructure construction is a complex system, it naturally exhibits the complexity associated with complex systems, as previously mentioned in this section. Not only is mega infrastructure construction a type of complex artificial system that embodies the property of complexity, but so is the mega infrastructure construction-environment compound system, which integrates mega infrastructure construction with the surrounding socio-economic and natural environments.

As a result, a challenge must be addressed regarding how to effectively and completely convey that mega infrastructure construction is a form of a complex system that embraces complexity. The first point of reference is the principle of human cognition, which progresses from the concrete to the abstract and from perceptual to rational thought. As a result, human people will initially perceive the physical complexity of the mega infrastructure building, such as the physical complexity of the tangible construction, due to the mega infrastructure construction's magnitude and the construction entity's technique and surroundings. In other words, it encompasses the wisest and direct knowledge of mega infrastructure development at the level of a hard system consisting of hard resources for mega infrastructure construction. Then, based on scientific preconceptions of a system, people conceptualize the physical complexity of a mega infrastructure construction hard system and thus express, through discourse, a system science capable of extracting the essential properties of a complex mega infrastructure construction system, such as a highly open environment, a heterogeneous entity, close connections between components, multiple constraints, and the evolution and emergence of a construction system's design (Ledford 2015). According to this definition, a mega infrastructure construction's complexity reflects and abstracts its physical complexity inside the category of a complex system. It represents the physical form of the mega infrastructure construction's complexity in complex system space.

The Difficulty of Managing Mega Infrastructure Construction

Because mega infrastructure building is an artificial complex system characterized by complexity, it is doomed to confront a complex management difficulty during its management operations. As a result, mega infrastructure construction management issues can be broadly classified according to the complexity of the construction itself or the construction environment. As seen in Fig. 1, because the construction and environmental complexity of the problems in region A (hence referred to as Type A) are relatively low, they are relatively straightforward problems that can typically be solved through mature experience and knowledge. Regarding Type B, obvious ambiguity and dynamic relationships will provide management challenges due to the high level of environmental complexity, and the inner structure of the construction is complex due to the high degree of construction complexity. Even in relatively simple construction environments, definite uncertainty and instability may develop. Additionally, due to the construction's strongly connected inner elements, implicit conduction and evolution of mutual impact between elements are possible. As a result, management challenges classified as Types B and C may typically be resolved by developing management norms, improving normal operating conditions, and leveraging mature experience and expertise.

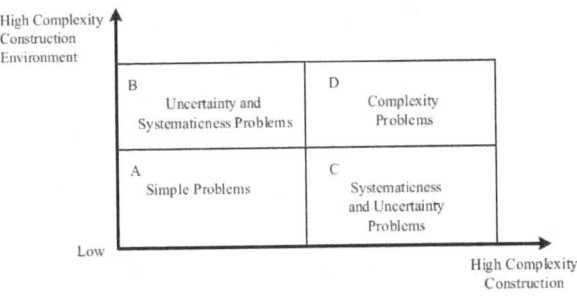

Fig. 1 Mega infrastructure construction management problem categorizations

Additionally, system development techniques can resolve problems that exhibit systematic behavior. This demonstrates that many management issues associated with mega infrastructure building (Types A, B, and C) can be overcome by combining scientific management and systematic management of mega infrastructure construction.

In terms of Type D, these are the types of complex problems that arise due to the high complexity of both the construction and the surrounding environment and must be resolved on a regular and efficient basis through the use of systematic complexity management ideas. Among these issues is the design of a heterogeneous organization management platform, an iterative pattern generation method for highly uncertain project decision making and planning, risk analysis and management for mega infrastructure construction, multi-subject coordination on a construction site, multi-objective comprehensive control, and the innovation of key construction techniques.

According to the analysis above, mega infrastructure construction difficulties can be classified into three categories. Combining the second level of Fig. 2, which contains Types A, B, and C, with the top level of Type D, a comprehensive system of mega infrastructure building management problems is produced (see Fig. 2).

The mega infrastructure development problem approach is commonly accepted to classify difficulties into three categories: complex management challenges, reasonably complex problems, and relatively easy problems. According to this classification, there are undoubtedly objective causes relating to the physical and systematic characteristics of the problems themselves; nevertheless, it also has a great deal to do with the management subject's cognitive ability. As a result, it cannot be concluded that any major infrastructure construction possesses a fixed and solitary problem system structure.

Rather than that, it is recognized that massive infrastructure construction is, to a degree, fluid and dynamic. For example, if two management subjects differ inability, the one with better ability will decide that there are fewer management challenges, while the one with less ability will disagree. Even if only one management subject is included, that individual will believe that complex problems at the top will diminish in complexity when information about mega infrastructure construction increases and the subject's cognition is promoted. The mega infrastructure construction cognition of those management subjects who are highly professional and have a wealth of expertise will assume that management problems are uncommon, implying that there are only two degrees of problems for them, rather than three.

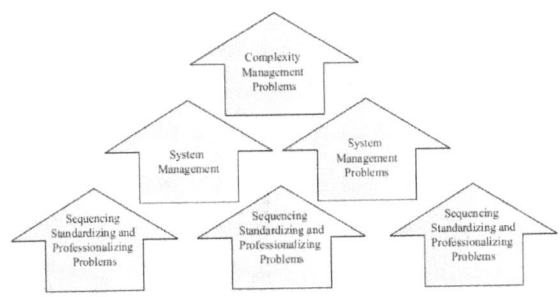

Fig. 2 Mega infrastructure construction management problem system

Additionally, mega infrastructure construction management is a systematic process to resolve complex management problems. It must encompass the three functions of complex management problem recognition, coordination, and execution. Thus, a mega infrastructure construction management system's fundamental structure is often made of the following three subsystems:

1. A mega infrastructure construction management recognition system whose primary job is to discover and analyze the construction complexity and systematic complexity of mega infrastructure construction and the complexity of mega infrastructure construction management difficulties.

2. A management coordination system for mega infrastructure construction whose primary functions are to design and decompose the complexity of management problems through the operation mechanism and processes of management organization and conduct a series of unique management strategies based on adaptability multi-scale management.

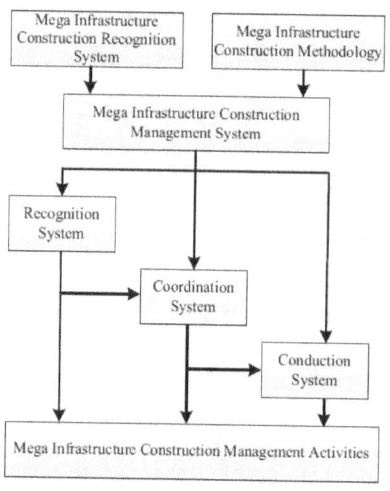

Fig. 3 Basic structure of mega infrastructure construction management activity

3. A management system for mega infrastructure construction whose primary role is to perform multi-subject coordination and comprehensive control of the construction site at all stages and levels of management by the coordination system's management objectives and strategies (see Fig. 3).

To summarize, mega infrastructure construction management activities, similar to mega infrastructure construction, are defined by their primary objective of analyzing and resolving complex problems during mega infrastructure construction creation. They are classified as a specific type of complex system, namely the mega infrastructure construction soft system, and exhibit the essential characteristics of complexity.

As a result, the evolutionary principle from systematic to complex construction management is formed by construction management to mega infrastructure construction management. The essential tenet is that managing massive infrastructure building is a complex system. The following are the primary functions of massive infrastructure construction activities:

1. To effectively organize and control the subject's self-learning and self-adaptation behaviors and coordinate work.
2. To develop a model of construction organization that incorporates self-organizational evolution, the flexibility of organizational elements, structure, and function, as well as the dynamic change of the construction environment and tasks.
3. To continually compress and synthesize the mega infrastructure construction objectives
4. To develop management solutions through a comparison, trial and error, iteration, and approximation process.

Through the self-adaptability and self-organization of the management subject, mega infrastructure construction management activities sustain and manage the complexity of mega infrastructure construction management difficulties.

In the domain of system science, the transition from systematic to complex does not involve:

. An increase in the quantity of systematicity but rather

an increase in quality.
- Resulting in a series of significant differences in management thinking principles.
- Methodologies of construction management and mega infrastructure construction management.

As a result, general systematic construction management approaches cannot be directly applied to mega infrastructure construction management or used to tackle complicated challenges within the mega infrastructure construction management problem system.

Additionally, mega infrastructure construction management problems encompass general systematicity and complexity, which does not necessarily imply increased systematicity. Rather than that, it is dependent on the quality of the shift in systematicity and the subsequent growth in complexity.

In this sense, complexity becomes a fixed property of problems involving the management of mega infrastructure construction. Given that major infrastructure building management activities involve integrating hard system and soft system components, the complexity is inevitably a composite of the hard and soft system components.

More broadly, the transition from systematic to complex transforms essential qualities associated with construction management to mega infrastructure construction management. This conclusion contains fundamentally important significance for establishing theoretical thinking concepts and developing theoretical systems for managing mega infrastructure projects.

Complex Integrity in the Management of Mega Infrastructure Construction.

Although the discourse system of system science has determined that complexity is an essential property of mega infrastructure construction management, the additional analysis should be conducted on the property's deeper connotation, thereby combining the characteristics of mega infrastructure construction creation activities.

The primary goal of any mega infrastructure project is to develop and plan a comprehensive artificial complicated system. In this context, integrity is the bedrock of major infrastructure development and management activities, comprising the physical and functional forms of the construction system and the entire system and process of associated management operations. In other words, all construction and management activities must be comprehensive and dependable to achieve the goal of creating an artificial construction system.

This means that, with the goal of human creation, not only must the hard construction system be integrated, but also the mega infrastructure construction management activities, which integrate their hard and soft systems and ensure the order and efficiency of mega infrastructure construction creation activities, must be integrated (Altshuler et al. 2003; Yu and Liu 2002). Otherwise, it will be impossible to achieve a completely artificial construction hard system.

Additionally, both generic and complex systems exhibit integrity from the perspective of theoretical logic. The issue is that when construction management progresses from systematic to complex, major changes in performance and the procedures used to achieve integrity occur. More precisely, the integrity of construction and its management activities might be compromised by reductionism and the composition of each sub-activity (Plotch 2015).

For example, in construction management activities, management objectives are embodied at the direct, visible, and physical level of the project, allowing the integrity management objectives to be reduced to the management objectives of each component and realized through the superposition of relatively simple methods.

However, integrity cannot be accomplished through such simplistic notions of reductionism and superposition in the case of massive infrastructure construction and management activities. For instance, the management objectives for mega infrastructure construction activities encompass those on the direct, visible, physical level and those on the indirect, unnoticeable, socioeconomic level. They encompass those of the same dimension and scale and those of varying dimensions and scales. Several objectives derive directly from the massive infrastructure construction-environment compound system. Under these conditions, it is exceedingly difficult to achieve the integrity of management objectives using basic reduction and superposition theories. This demonstrates that a type of complex integrity manifests itself in mega infrastructure construction management activities through the overall design of management objectives and a variety of other management problems that reflect the complexity of mega infrastructure construction management activities (Flyvbjerg 2003).

Thus, while integrity is a shared characteristic of general and mega infrastructure construction, the former encompasses the superposition of integrity in general systems, referred to as general integrity. In contrast, the latter encompasses the non-superposition of integrity in complex systems, referred to as complex integrity. Thus, when a mega infrastructure construction management system combines a hard system for construction and a soft system for management to form the entirety of activity, the complexity of the mega infrastructure construction complex hard system (management target system) and complex soft system (management subject system) is combined to create the complex integrity of the mega infrastructure construction management activities.

In this regard, it is believed that the origins and deep meaning of the complexity of mega infrastructure construction management activities stem from the building of mega infrastructure and complicated integrity. Thus, within the mega infrastructure development management activities, it is necessary to establish concepts based on the analysis and resolution of complicated integrity problems and methodologies for researching complex integrity problems.

THE GOOD, THE BAD, AND THE BEST OF MEGAPROJECTS

While undertaking large infrastructure projects is always risky, there are strategies to increase the likelihood of a successful landing. Megaprojects in the infrastructure sector are critical for the future of cities, states, and individual livelihoods. The issue is that these projects frequently run over budget or schedule — or both.

It's vital to remember, though, that developing and maintaining infrastructure is a critical and occasionally life-saving endeavor. For example, sewage and water delivery systems help keep diseases like cholera at bay. Much of the Netherlands would be submerged without the North Sea Protection Works.

Large-scale infrastructure projects can also have a revolutionary effect on the economy. Take the Panama Canal as an example. It contributes a sizable portion of the country's GDP. Dubai's international airport is the busiest globally, accounting for 21% of employment and 27% of GDP. And Hong Kong would undoubtedly come to a grinding halt without its clean and efficient subway system, the MTR, which has enabled the densely populated city to expand outside its core districts.

These mega projects have been successful; it is nearly impossible to imagine these locations without them. Unsurprisingly, they are all being extended. Not unexpectedly, the canal and airport projects are both behind schedule and significantly over budget; even the ultra-efficient MTR has encountered delays with some of its projects.

McKinsey estimates that the world will need to invest approximately $57 trillion in infrastructure by 2030 to support the projected levels of global GDP growth. Around two-thirds of it will be required in developing economies, where middle classes are growing, the population is growing, urbanization is increasing, and economic growth is increasing. These countries require infrastructure, yet far too frequently, years pass without the promised road, bridge, or metro projects materializing.

The dangers inherent in megaprojects—those costing more than $1 billion—are widely established. Bent Flyvbjerg, a specialist in project management at Oxford's business school, claimed that nine out of ten projects went over budget in one significant research. Rail projects, for example, consistently exceed budget by 44.7 percent on average, and demand is overstated by 51.4 percent. According to McKinsey, bridges and tunnels incur an average cost overrun of 35%; highways incur 20%. Many projects are approved with a 20% estimated return on investment, so governments are left to foot the bill for the remainder.

Time overruns are also a constant issue. Consider Salvador, Brazil's metro system. Although construction began in 2000, it took more than a decade for the first passengers to board; most of the work is still incomplete. It took a decade for New York to begin the $3.9 billion project to repair the 59-year-old Tappan Zee Bridge; in the meantime, the cost of maintaining the deteriorating bridge continues to rise.

Finally, the idea that initiatives must function on two levels —short-term financial recovery and long-term social impact—often becomes a barrier to action. Even necessary projects are not always completed, particularly in areas where the cash generated by the project is unlikely to cover its cost.

How can businesses and governments improve their ability to deliver megaprojects on time and budget while also delivering societal benefits? To address that topic, we examine the primary reasons why megaprojects fail and then make recommendations for improvement.

Why do projects go bad?

We can identify three primary reasons for failure.

The first is, excessive optimism and complexity. To justify a project, costs and schedules are routinely underestimated, and benefits are routinely overstated. Flyvbjerg contends that project managers vying for funding manipulate data until they fall inside the bounds of what is deemed affordable; stating the true cost, he writes, would render the project undesirable. Such endeavors are doomed from the start.

This is frequent when large projects span state or national borders and involve a mix of private and public funding. For instance, a new railway may involve three national governments, multiple municipal governments, varying environmental and health requirements, varying levels of skill and salary expectations, and dozens of private contractors, suppliers, and end-users. A single issue can cause the process to delay indefinitely. In one instance, it took a decade for two countries to iron out the diplomatic details necessary to build a hydropower project. All too frequently, these complicated factors are not thoroughly evaluated or fully priced before initiating a project.

A useful reality check compares the current project to similar finished initiatives. This technique, dubbed "reference-class forecasting," combats confirmation bias by compelling decision-makers to explore scenarios that do not always support the favored course of action. For instance, if a city wishes to construct a ten-kilometer metro line with four stations, it should consult with other towns that have constructed comparable lines to understand the true cost and time dynamics better.

Second would be ineffective execution. After delivering an unrealistically low project budget, the temptation is to cut corners to maintain cost assumptions and protect the engineering and construction firms contracted to deliver the project's profit margins. From design and planning through construction, the execution of a project is fraught with complications, including insufficient design, a lack of defined scope, ill-advised shortcuts, and even mathematical errors in scheduling and risk assessment. A McKinsey assessment of 48 failed megaprojects found that in 73% of cases, cost and time overruns resulted from poor execution; the remainder were the result of politics, such as new administrations or regulations. Part of the reason for poor execution is that many projects are so complex that what appear to be minor concerns can quickly escalate into huge headaches. For instance, if steel does not arrive on schedule at the job site, the delay can cause the entire project to stall, likewise if one of the specialized trades encounters difficulties. Increased productivity will not compensate for these shortcomings, as delays reverberate throughout the project system.

Third issue is a lack of productivity. While manufacturing productivity has nearly doubled in the last two decades, construction productivity has stayed steady or even decreased. In many markets, wages, on the other hand, have continued to climb faster than inflation, resulting in greater expenses for the same outputs.

According to McKinsey research, delivering infrastructure more effectively can result in a 15% reduction in total cost. Preparation in advance pays for itself many times over. In aggregate, efficiency advantages in approval, engineering, procurement, and construction can result in savings of up to 25% on new projects without sacrificing quality.

Organizational design and capability deficiencies. Numerous organizations involved in megaproject development have an organizational structure in which the project director is four or five tiers below the top leadership. The following structure is frequently used:

- Layer 1: Subcontractor to the contractor
- Layer 2: Contractors to a construction manager or managing contractor
- Layer 3: Construction manager to owner's representative
- Layer 4: Owner's representative to project sponsor
- Layer 5: Project sponsor to business executive

This is a concern since each layer will have an opinion on reducing time and expenses. For instance, the first three layers are motivated by additional work and money. In contrast, the subsequent layers are motivated to complete projects on time and within budget. Additionally, the authority to make final choices is frequently dispersed across the activity.

Capabilities, or the lack thereof, are another point of contention. Large-scale projects are often funded by the government or by an entrepreneur with audacious ambitions; they can take ten to fifteen years to complete. Even individuals who make their profession building huge infrastructure projects may complete only three or four mega projects in their lifetime. Because each one is unique, the learning curve is steep, and the necessary skills are limited. These issues are exacerbated by the haste with which initiatives are initiated. Megaprojects may be required to establish organizations of several thousand people from scratch within 12 to 18 months—a substantial operational and managerial challenge comparable to establishing a new start-up company.

How to keep megaprojects on track?

Any large project entails a high risk of failure, but consistently running late and over budget shows that systematic errors occur. This means that these issues can be detected and resolved.

Typically, when initiatives fail, hindsight reveals that the difficulties began at the onset due to inadequate justification and need for the project, stakeholder misalignment, insufficient planning, and an inability to locate or use the necessary capabilities. Costs are frequently underestimated, while advantages are exaggerated. As a result, the baseline for evaluating the project's overall performance is incorrect. The critical point is to set social and economic priorities first and then determine which projects are best suited to achieving them. This necessitates the development of robust and independent analyses of the genuine costs and benefits. Certain countries are closer than others to realizing this aim. Singapore, for example, has a national objective of dense urban living, with public transportation accounting for 75% of trips. This aim informs the Land Transport Authority's transportation project selection process. Without such vigilance and monitoring, one can envision bridges that lead nowhere, oversupplied power, and deserted highways.

To ensure accountability, the process for selecting projects must be fact-based and transparent. South Korea exemplifies one promising strategy. It established the Public and Private Infrastructure Investment Management Center (PIMAC) in 2005 to obtain accurate cost and benefit data; PIMAC conducts feasibility studies on public megaprojects, conducts value-for-money analyses, sponsors comprehensive research on improving public investment, and evaluates completed projects. PIMAC has rejected nearly half of the projects it has assessed thus far. Before PIMAC, the rejection rate was 3%.

Another characteristic of distressed projects is that they lack proper controls. They lack thorough risk analysis and risk management methods and timely reporting on progress against budgets and timeframes. The data used to report on project progress is frequently out of date (since it is typically based on contractor payments rather than actual work accomplished) and out of step with the true development of the project. Additionally, baselines are modified regularly, and contractors and owners track progress using various measures. It becomes troublesome when numerous estimates of the project's cost and time performance relative to the baseline exist, implying no shared understanding of performance. This constrains the partners' capacity to devise strategies for expediting project execution and limiting cost overruns.

A more advanced technique would be to use real-time data to compare field activity, such as cubic meters of concrete poured or soil moved, to work plans and budgets. This is in contrast to the conventional technique, which monitors progress by comparing cash paid to contractors to the budget. Measuring progress based on cash flow, on the other hand, is less than ideal, as payment typically takes longer than 30 days. This means that the underlying data is outdated; contractor payments may not correlate with real construction progress. Enhancing project performance necessitates improved planning and preparation in three key areas.

Conduct engineering and risk assessments before initiating construction. This is frequently stated but rarely implemented, even though it enhances project performance. Edward Merrow, the founder of the Independent Project Analysis firm and author of a book on megaprojects, has demonstrated that the greatest examples of project definition work result in a roughly 20% reduction in both project deadlines and costs.

For three reasons, most project development organizations and sponsors are hesitant to invest large funds in early-stage engineering and design. First, they frequently lack the cash necessary to invest considerably in design and engineering at the early stages of a project. Second, they are eager to begin building and breaking ground. Finally, they are concerned that the design will be altered during construction, rendering the expense of the up-front design worthless.

Our experience indicates that investing 3 to 5% of the project's capital cost on early-stage engineering and design produces much better results in on-time and on-budget delivery. This is because the design phase frequently uncovers issues that must be rectified before the commencement of construction, thereby saving time and money.

Simplify the permitting and land acquisition processes. It is not uncommon for a project to take longer to obtain permissions than to develop. Permitting best practices in-

cludes:
- Prioritizing projects.
- Creating clear roles and duties.
- Establishing time constraints throughout the process, particularly during public review.

Providing a "one-stop-shop" for permission might be beneficial. England and Wales reduced the time required to approve power-industry infrastructure from 12 to 9 months by using these measures, compared to an average of four years in the rest of Europe.

Additionally, projects might be structured to eliminate time-consuming land battles. Virginia finalized a plan to widen Interstate 495 in 2012 after a private design firm proposed a concept that significantly reduced costs and avoided the need to demolish hundreds of residences.

Create a project team that is well-balanced in terms of abilities. Without a well-resourced and qualified network of project managers, counselors, and controllers, projects will fail to achieve the maximum return on investment achievable. They will, at the very least, fail.

Investors and owners must be actively involved in assembling the project team. It is insufficient for them to have a hazy theoretical understanding of how the project should operate. They must develop a comprehensive, practical plan to address such potential occurrences as managing quality concerns, increasing contractor expenses, or replacing a high-tech supplier. A skilled project manager alone is insufficient; participants must construct a team that possesses all necessary talents, including legal and technical expertise, contract management, project reporting, regulatory approval, stakeholder management, and government and community relations.

The world requires megaprojects to offer billions of people

economic and social benefits and generate the economic development necessary to pay for them. However, a failed project has far-reaching consequences that extend beyond a single bridge, tunnel, or sewage system. Getting it right, or at the very least, better, benefits everyone.

MAKING MEGAPROJECTS MODULAR

As a result of climate change, many sectors consider significant modifications to technology and basic infrastructure. The previous century's oil and coal-fired power plants are being phased out in favor of wind farms and solar arrays. Automobiles and gas station networks powered by fossil fuels may soon be a thing of the past. Almost every industry will require significant capital investments, which will entail significant risks.

I've spent the last three decades researching and consulting on megaprojects. I've discovered two important criteria in determining whether an organization succeeds or fails: reproducible modularity in design and rapid iteration. If a project can be completed quickly and in a modular fashion, allowing for experimentation and learning along the way, it has a good chance of success. It is almost certain to have difficulties or fail if it is attempted on a large scale using one-off, highly integrated components.

Regrettably, the standard for conventional commercial and government megaprojects — such as hydropower dams, chemical processing plants, aircraft, or big-bang enterprise resource planning systems — is still to develop something monolithic and bespoke. Before such projects produce benefits, they must be entirely complete: Even when a nuclear reactor is 95 percent constructed, it is useless. Components are frequently customized, with a high degree of uniqueness rather than modularity, limiting opportunities for learning and raising the costs of integration and rework when issues develop. New technology and bespoke designs are becoming more prevalent, impeding speed and modular scale-up. Additionally, the size of megaprojects is often determined many years in advance of the start of activities. This is a recipe for catastrophe if more capacity is provided than is eventually required or if demand exceeds expectations and more capacity cannot be added. The Channel rail tunnel connecting France and the United Kingdom, for example, has a fixed capacity, and because tunnel utilization is only half of what was planned, vast and expensive capacity remains unutilized. The investment has proven to be a financial disaster.

Cost overruns may be irrelevant if you are a huge multinational corporation like BP or Tesla considering a $10 million project. Coming in $10 million over budget would have little impact on the bottom line at such firms. However, when the budget forecast begins at $10 billion, the stakes are far higher, even for governments. As a result, smart organizations employ processes and technology that facilitate modularity and rapid learning and needlessly complicated redo when issues develop.

Much of this will sound familiar and sensible to entrepreneurs in the technology business. However, huge corporations and governments have not yet internalized these lessons regarding large-scale undertakings. While many megaprojects, such as bridges or power plants, are unlikely ever to be modular, there is still plenty of room for selecting technologies that permit quick scaling and incorporating modularity through the imaginative application of tried-and-true technology. Let us begin by examining the elements that contribute to projects scaling up quickly.

Why Speed and Modularity Matter

Speed is critical to the success of megaprojects, as extended timetables introduce additional risk and uncertainty. Philip Tetlock, a professor at the Wharton School, has proven that people can foresee some events with reasonable accuracy overtime periods of up to one year. These events include GDP growth, macroeconomic policies, business cycles, technology advancements, and geopolitical wars. After then, the accuracy rapidly drops, and beyond a three- to five-year time horizon, it vanishes into the mists of randomness.

And Tetlock's evaluation is probably unduly optimistic. His findings are based on the work of highly talented forecasters, and the projections he examines are simplified, frequently expressed as yes or no responses to queries such as "Will any country leave the eurozone in the next year?" or "Will North Korea launch a nuclear weapon by year's end?" Most real-world projections do not have binary outcomes but rather encompass a wide range of probable outcomes. They cover topics such as "How many people are anticipated to die in the next year from Covid-19?" and "How much is the California high-speed rail system going to cost?" Although binary questions are easier to answer than multiple-choice questions, the latter is more popular in practice.

Entrepreneurs and financiers in Silicon Valley have long recognized the vital nature of speed in winner-take-all markets. In the technology business, new enterprises place a premium on generating a minimum viable product during the first year and attaining a market leadership position by three to five years. Reid Hoffman, the co-founder of LinkedIn, refers to this process as "blitzscaling" and contends that scale-ups, not start-ups, differentiate Silicon Valley from other digital ecosystems.

The speed factor is only half of the equation. Eric Schmidt, former chairman and CEO of Alphabet, and Jonathan Rosenberg, former senior vice president of products at Google, identify the other half: Iterate and ship. "Build a product, ship it, monitor its performance, design and implement improvements, and re-release it," they say. "The companies that complete this process the quickest will win."

Iteration ensures that the quality of the delivered product continues to improve over time. As Harvard Business School professors' emeriti Carliss Baldwin and Kim Clark demonstrated more than two decades ago, iteration enables learning by establishing a feedback loop. The experience gained from delivering one module improves the experience gained from delivering the next repeatedly. Additionally, iteration allows for experimentation. Rather than quickly scaling up, you experiment with a few modules, refine the next, and repeat until you master delivery, at which point you scale up. It's clear to see how speed plays a role in this process—the faster you iterate, the more you learn, and the farther you can reduce costs while increasing safety and efficiency.

Humans are inherently adept at testing and learning, which is why a company built on modular replicability has a greater chance of success than one built on long-term planning and forecasting, which humans are notoriously awful at.

Next, let's examine a megaproject that exemplifies a prudent scale-up.

Giga Nevada: Smart Scaling

Tesla's Gigafactory 1, dubbed Giga Nevada, is a $5 billion high-tech lithium-ion battery manufacturing facility under development east of Reno. The megaproject's objective is to reduce the cost of electric vehicles and household energy systems by manufacturing batteries on a massive scale. If all goes according to plan, Gigafactory 1 will have the world's largest footprint, covering more than half a million square meters or 107 football fields.

The structure is designed in a modular fashion. Tesla began by defining a minimum viable production facility, or "block," that could be operational immediately upon completion, imparting learning as further blocks were constructed. Gigafactory 1 construction began in late 2014, and by the third quarter of 2015, the first phase was complete, producing the Tesla Powerwall, a residential energy storage system. Tesla celebrated the factory's grand opening in July 2016, with three of the factory's 21 blocks completed, accounting for around 14% of the entire projected size. Battery cell mass manufacturing commenced in January 2017, a little over two years after the project's inception. This is significantly faster than is customary for projects of this magnitude when operations typically begin five to seven years after construction is complete. In 2014, Gigafactory 1 was anticipated to have a capacity of 35 gigawatt-hours per year. That capacity seems to have been achieved even before the completion of the factory, indicating that significant learning in construction and manufacturing has taken place.

Tesla earned two significant benefits as a result of its emphasis on speed. To begin, the corporation mitigated the risk of cost overruns, which tend to grow as project timetables drag on. Second, it began earning revenue within a year of deciding to proceed with the project—significantly sooner than would have been the case if it had used the typical approach to megaprojects. Both benefits are critical for rapidly growing businesses that cannot afford capital locked up in lengthy, hazardous development projects. Regrettably, traditional megaprojects frequently fail miserably.

Monju and the Problem of Negative Learning

Japan's Monju nuclear power facility was the world's first commercial fast-breeder reactor. It was supposed to serve as the cornerstone of a high-priority national initiative to reuse and eventually generate nuclear fuel in a country with few indigenous energy sources.

The facility was customized: each element and component were developed and manufactured specifically for a specific function, utilizing cutting-edge technology. Construction began in 1986, and initial criticality (a sustained fission chain reaction) was achieved on time eight years later, in 1994. After that, test operations commenced, followed by an August 1995 inauguration. In December of that year, a severe fire forced the facility's closure, resulting in a five-year delay that was significantly prolonged as other issues were discovered. Test runs began in 2010, and shortly after that, a three-ton refueling mechanism toppled into the reactor vessel. Recovering the machine took nearly a year.

Further issues and the revelation of major maintenance deficiencies prompted Monju to postpone preparations for restarting the reactor for commercial usage in May 2013. The Nuclear Regulation Authority declared Monju's operator unqualified to operate the reactor, and the government permanently shuttered the plant in December 2016.

Monju is believed to have generated electricity for only one hour over its 22-year lifespan while spending more than

$30 billion and $12 billion. Decommissioning will take an additional 30 years, until 2047, for $3.4 billion. If past practice is any indication, those figures are excessively optimistic, with future delays and cost overruns a near certainty. Monju will, at the very least, end up as a 60-year, $15 billion endeavors with no or negligible advantages. Monju is not the only example—it is only one of the most conspicuous.

The contrast with Tesla could hardly be more glaring. Monju's design had nothing in common with the replicable production modules at Gigafactory 1, where learning was continuous and scaling became better and better and faster. At Monju, everything was completed in a single step, despite its enormous complexity. This resulted in a condition known as negative learning, in which learning slows rather than accelerates progress. The more information the Monju team gathered, the more difficulties and further effort became apparent.

As with Monju, many megaprojects are difficult to break down into replicable units capable of rapid learning and improvement. When you dig a hole in the dirt, for instance, everything appears to become uncertain, bespoke, and slow. However, difficult does not imply impossibility. Almost any project can have significant portions of its work replicated, allowing even the least-scalable projects the opportunity to turn negative learning into positive learning. Scalability is not an either/or proposition: either it is scalable, or it is not. Scalability is a matter of degree: incorporating as much scalability as possible into any project, even the most improbable ones.

Consider the following example.

Madrid's Modular Metro

Manuel Melis Maynar is well aware of the value of scalability. As a seasoned civil engineer and president of Madrid Metro, he oversaw one of the world's largest and fastest subway extensions. Subway construction is widely regarded as being manual and slow by nature. It can take up to ten years from the time a decision to invest in a new line is made to the time trains begin operating, as was the case with Copenhagen's recent City Circle Line. And that's if no complications arise, in which case it could take 15 to 20 years, as was the case with London's Victoria line. Melis reasoned that there had to be a better approach, and he discovered one.

Begun in 1995, the Madrid subway extension was completed in two four-year phases (1995–1999: 56 kilometers of track, 37 stations; 1999–2003: 75 kilometers, 39 stations), owing to Melis's groundbreaking method to tunneling and station construction. In terms of project management stands in stark contrast to the Eurotunnel's record, which has been extremely costly to its investors. Melis's success was due to applying three fundamental guidelines to the project's design and administration.

No monuments.
Melis chose against using signature architecture in the stations, even though such ornamentation is customary, with each station frequently created as a separate monument. (Consider Stockholm, Moscow, Naples, and the Jubilee line in London.) Melis was well aware that signature architecture is known for delays and cost overruns, so why invite trouble? His stations would all be modular in design and constructed using proven cut-and-cover techniques, allowing for replication and learning as the metro progressed.

No new technology
The project would forego novel building techniques, train

carriages, and designs. Again, this approach runs counter to most subway planners, who frequently take satisfaction in delivering cutting-edge signaling systems, driverless trains, and so on. Melis was well aware that new product development is one of the riskiest endeavors any firm, including his own, can undertake. He was not interested in any of it. He was only concerned with what worked and could be accomplished quickly, inexpensively, safely, and with a high level of quality. He integrated tried-and-true items and methods in novel ways. Does that ring a bell? It ought to. This is how Apple innovates, and with great success.

Speed

Melis saw that time is similar to a window. The larger it is, the awful stuff that can pass through it, including unanticipated catastrophic catastrophes known as black swans. He deliberated for a long time on drastically reducing the size of his window by reorganizing tunneling work for speed. Historically, communities constructing a metro system would hire one or two tunnel boring machines to complete the operation. Melis instead calculated the optimal length of tunnel that a single boring machine and the team could deliver—typically three to six kilometers in 200 to 400 days—divided that length by the total length of the tunnel he required and then hired the appropriate number of machines and teams to meet the schedule. He utilized up to six computers simultaneously at times, unheard of when he initially did it. His module unit was the optimal tunnel length for a single machine, and like the station modules, the tunnel modules were repeated indefinitely, encouraging positive learning.

Unexpectedly, the tunnel-boring teams began competing

against one another, pushing the pace even further. They'd convene at night in Madrid's tapas bars to share notes on daily progress, ensuring their team was ahead of schedule and imparting knowledge in the process. Additionally, by having multiple machines and teams operating concurrently, Melis could systematically assess which machines and teams performed the best and hire them the next time around. Additional positive learning. A feedback mechanism was established to minimize time-consuming arguments with neighborhood organizations. Melis persuaded them to accept tunneling 24 hours a day, seven days a week, by publicly asking whether they wanted a three-year or eight-year tunnel building timeframe.

There will be no monuments, no invention, and everything will be modular and quick. That sounds like a recipe for uninteresting, low-quality design. However, in Madrid, you'll discover huge, functional, airy stations and trains—a far cry from London's and New York's dismal, confined tunnels. Melis's metro is a workhorse devoid of glitzy technology that could jeopardize operations. It ferries millions of passengers every day, year after year, precisely as designed. Melis accomplished this at a cost that is half that of industry averages and at a twice as fast rate—something most believed was unachievable.

A Wiser Path

The disparate experiences of the main initiatives discussed here imply that when firms and governments begin on large-scale endeavors, they must carefully select and prudently invest in technologies that lend themselves to smart scaling.

Consider the energy industry once more. To survive, it must break the vicious circle of negative learning now in place and decipher the code of rapid, replicable scale-up. Small modular reactors (SMRs)—nuclear power plants with an estimated cost of $1 billion per unit—aim to accomplish precisely that. The projected building of an SMR in Wyoming, financed by Bill Gates and Warren Buffett, may represent the first step in that direction. However, with a seven-year projected completion date, the project is still moving at a snail's pace. We do not have time to wait, given the impending climate crisis.

While projects have supplanted operations as the economic engine of the modern economy, many executives continue to undervalue project management. Businesses must reimagine their strategies, including a new set of tools and principles.

The wind is a more scalable alternative. By their very nature, Turbines are modular and replicable, making them perfect candidates for intelligent scale-up. They were initially built on-site, but the fledgling business rapidly realized this was wasteful and changed to make them indoors, utilizing industrial processes and logistics that could be successfully managed and optimized. When the London Array in the United Kingdom was finished in 2013, it was the world's largest offshore wind farm, costing $3 billion in 2012 dollars. The project started development in March 2011 and produced power in October 2012. All turbines were completely operational by April 2013, just two years and one month after construction began. And, less than a decade later, that is no longer considered fast. In 2018, the Walney Wind Farm addition was completed less than a year. It is located off the coast of England and features 87 turbines.

Not just the energy sector is moving away from the classic megaproject. Take the space business as an example.

NASA's complex designs often take a decade to plan and another decade to build. Its missions are too large to fail and too slow to restart in the event of a failure. The longer the period, the greater the danger of ultimate failure, with little opportunities for learning along the way. However, a new breed of space entrepreneurs (including Elon Musk) significantly reducing costs and delivery times by relying on conventional, industrially built building blocks.

Consider the instance of Will Marshall, a young engineer who began his career at NASA's Jet Propulsion Laboratory. He eventually became tired of Big Space's sluggishness and waste and resolved to do things differently. He co-founded Planet Labs with two other NASA alumni and created a satellite dubbed Dove in his Cupertino, California, garage.

With a weight of ten pounds, a production period of a few months, and a cost of less than $1 million (including launch and operations), Dove satellites are far smaller, faster, and less expensive to manufacture than anything NASA has ever attempted—yet they are equally highly built and more agile. Each satellite is composed of three CubeSat modules, each of which comprises multiples of 10x10x10 cm modules, or what Marshall refers to as Legos. The CubeSats' electronics and structure are made from commercial off-the-shelf components, similar to those used in cell phones and recreational drones, which keeps costs and delivery times low. Planet Labs launched several hundred satellites in the 2010s—the largest constellation ever launched—giving real-time data for climate monitoring, agriculture, disaster response, and urban planning.

In 2014, Planet Labs lost 26 Dove satellites. They were perched atop a massive rocket that detonated on the launchpad. Compared to his nine successful launches at the time, the loss had a negligible effect on the firm. Satellites that were lost were swiftly replaced, and new ones were launched into orbit. Marshall's modular strategy ensures that each mission is inexpensive to fail and quick to reproduce in the case of failure, with lessons learned instantly applied in the subsequent iteration.

My suggestion to anyone considering a large project is to follow Tesla, Planet Labs, Madrid, and wind farms as examples. Wherever possible, choose fundamental technologies that lend themselves to modularity and replication. Where this is challenging, apply tried-and-true technology in novel, modular ways so that you may learn as you go, lowering costs and increasing speed with each iteration. If this strategy is effective for something as difficult and customized as excavating a subway beneath a city, it is likely useful for virtually every job. The possibilities are as rich as your imagination.

8 PLANNING AND ORGANIZING BEST PRACTICES FOR MEGAPROJECT SUCCESS

Megaprojects in construction live up to their reputations in various ways, including their scale, expense, complexity, and risk. Africa is endowed with an abundance of natural resources, which, when combined with rapidly rising economies, results in a continent with limitless industrialization potential. In the past, a lack of adequate infrastructures such as railways and roads hampered any significant development on the continent. Still, things are looking up with greater investment and interest from international parties such as China.

We next shift our focus to the first phase of megaproject management, delving into the eight best practices for early planning and organization. We believe that the planning and organization establish the stage or blueprint for everything that occurs after the project's approval. During planning and organization, you establish the project's course —and make course corrections as necessary. Once equipment, supplies, and personnel are included, much planning flexibility is lost.

#1 – Assign the project team as soon as possible

Assign members of the project team immediately. Too frequently, building owners make the error of utilizing separate crews for each project step (e.g., concept, pre-feasibility, feasibility, execution, start-up, and turnover). To ensure accountability, transparency, and responsibility, it is preferable to appoint a core team that will remain involved throughout the project.

#2 – Select the most appropriate project delivery method

Once the project has progressed through strategic planning and into pre-feasibility, the team must settle on a specific project delivery approach. The delivery strategy chosen will influence the project's cost, timeline, design quality, construction technique, and long-term maintenance requirements.

#3 – Create reasonable estimates

When creating project estimates, project teams must be prudent and realistic. Because CEOs are naturally optimistic about the projected benefits of a project, they might also be too optimistic about cost and schedule restrictions. Validate first concept estimates and uncover significant estimating errors and omissions with your core team. During the design development stage, rely on industry-recognized parametric cost estimates and benchmark data rather than on your internal estimates. Create pricing models that take into account a range of probable outcomes.

#4 – Actively manage project risks

Risk management is a popular buzzword in today's corporate world! Risk management is the process of recognizing potential threats to the project's effective completion and implementation, both internal and external. Early in the project's pre-feasibility and feasibility stages, risk planning, and identification are critical for developing a thorough and accurate estimate of the project's duration and cost. The project team is directly controllable and influenceable certain hazards, while others are completely out of their control. Effective stakeholder communication and risk management may help mitigate a wide variety of project hazards, both internal and external, to the project.

#5 – Secure senior management support

To be successful, every business megaproject requires strong top management support. Before the project can be approved, the project team must produce and vet a comprehensive project charter, execution plan, baseline budget, and baseline timeline. All engineering plans must be analyzed and developed to the point where the project can be imagined and built "on paper." Most significantly, the project must be strategically connected with the current aims and objectives of the firm and its stakeholders. The project must be relevant, and senior management must believe it is the best use of capital funds, and it must be planned to provide concrete benefits.

#6 – Specify policies and procedures for specific projects

As any seasoned project manager is well aware, the success of a megaproject is contingent upon the extended project team collaborating effectively. Collaboration involves a well-defined set of policies and processes, well-defined roles and duties, and frequent communication via approved means. While your organization may have policies, standards, and procedures for managing major capital projects, it is advised that the project team design customized rules and processes tailored to the mega-unique project's demands and conditions.

#7 – Assign roles and duties to specific projects

Your project team will perform optimally if each member is aware of their distinct tasks and responsibilities. As is the case in other businesses, what appears to be self-explanatory on the surface becomes more difficult when the pen meets the paper. Create an organizational chart that clearly and completely identifies all essential participants and third-party resources and the hierarchical reporting links.

#8 – Regularly convene team meetings
Your project team will perform optimally if each member is aware of their distinct tasks and responsibilities. As is the case in other businesses, what appears to be self-explanatory on the surface becomes more difficult when the pen meets the paper. Create an organizational chart that clearly and completely identifies all essential participants and third-party resources and the hierarchical reporting links.

Planning and organizing ahead of time will pay off.
You cannot manage a megaproject based on industry experience, project management concepts, and technological application. A successful megaproject is partially the product of great planning and organization. Africa presents its own set of difficulties to projects, and a thorough grasp of these complexities through excellent early-stage project management will significantly ease the smooth operation, hence lowering potential mega risk'.

THE CAUSES AND CURE FOR UNDERPERFORMING MEGAPROJECTS

Juliano Denicol, Andrew Davies, and Ilias Krystallis's new research, recently published in the Project Management Journal, examines "What Are the Causes and Cure for Suboptimal Megaproject Performance." It is based on a systematic evaluation of the literature in which 6,007 titles and abstracts and 86 full publications were analyzed. The authors found 18 causes and 54 remedies for underperforming megaprojects.

According to the report's preface, "megaprojects are extraordinarily risky endeavors that are notoriously difficult to manage and frequently fail to accomplish their initial objectives." That is why they are an intriguing topic for discussion at the Strategic Management Forum's next conference. "Unafraid: How Successful Leaders Overcome Insurmountable Obstacles and Avoid Predictable Surprises." It will take place on March 9th in London. And Dr. David Hancock, Director of the UK Government Cabinet Office Infrastructure and Major Projects Authority, is among the speakers.

The new analysis puts the magnitude of the situation into context by quoting studies indicating that between 2013 and 2030, US$57 trillion will be spent on infrastructure improvements (McKinsey Global Institute, 2013). The Economist (2008) stated that "the greatest investment boom in history is underway." As a result, we stand to lose a great deal if we do not enhance the results of megaprojects. On March 9th, the conference will feature speakers who will discuss Active Thinking, Improved Decision Making, Systems Thinking and Critical Systems Thinking, Value Creation, Living Business Models, Taming, Messy, and Wicked Problems, Risk Leadership, and Complexity. The report's authors' assertions and findings are particularly intriguing in this perspective.

"While enormous efforts have been made to increase our understanding of megaproject performance," they write, "each contribution sheds light on a particular or isolated event." This is because "there is no overarching theory or framework that can bring the disparate contributions together into a coherent picture illustrating how performance is contingent on various components — such as decision making, integration, leadership, and teamwork — cooperating as an integrated whole."

The authors of the paper assert that "consideration of their interdependence may inform discussions about how megaprojects can be more thoroughly studied to improve our understanding of topics such as value (co) creation, its evolution, scope, organizational boundaries, and transferability across the ecosystem."

They reference their own "recent call for research on new project delivery methods," implying that this "could be an interesting way of envisioning megaprojects as value creation and capture systems." (2017) (Davies et al., 2019). Additionally, the concept of "exploring megaprojects as dynamic inter-organizational systems — covering development, delivery, and operations — and identifying methodologies for designing the evolving system's architecture." (Denicol, forthcoming).

Juliano Denicol, Andrew Davies, and Ilias Krystallis conclude their article by "suggesting that future research and theory development should take a systematic approach, incorporating some of the several factors affecting megaproject performance." Additionally, they note that "the literature could be enriched by research that views megaprojects as production systems and by investigating their separate components via a systems lens."

THE INCREASING DIFFICULTY OF BUILDING MEGA-INFRASTRUCTURE

While the cognitive principles underlying complex systems and the complex entirety of mega infrastructure construction and management have been established, the types of complex changes that will occur within the scope of complexity in future mega infrastructure construction and management must be identified.

The whole complexity of planned mega infrastructure projects can be represented in two dimensions: technological complexity and managerial complexity.

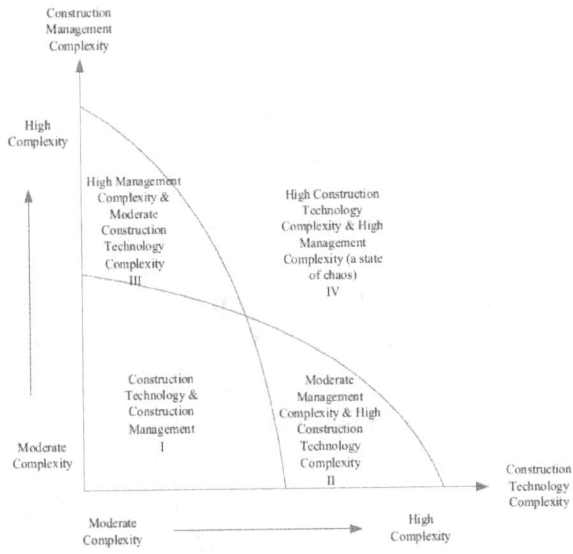

Fig. 4 The overall complexity of mega infrastructure construction

As seen in Fig. 4, mega infrastructure construction is moderately complex in terms of construction technology and construction management in Region I. In contrast, it is moderately complex in construction technology and construction management in Region II. Due to the complexity of construction technology, materials, equipment, and procedures for mega infrastructure building are not yet complete, and construction technology's necessary principles and laws are not evident. A moderate level of technological complexity characterizes mega infrastructure construction in Region III but a high level of management complexity. Due to the complexity of management, it is difficult to perceive, coordinate, and implement the complexity of mega infrastructure building management and efficiently and orderly control the complex totality of management. Finally, in Region IV, mega infrastructure construction manifests as a high level of technological construction complexity and a high level of management complexity, indicating that the degree of complexity is in a state of anarchy on an aggregate level.

Additionally, the great overall complexity of mega infrastructure construction is only a conceptualization of the whole, i.e., a hazy image. It is imprecise because new cognitions and conceptions of the complex system and its complexity will emerge in the future. Due to tremendous advances in human intelligence and capacities, what is considered complicated will become less complex in the future. However, if complex concepts are comprehended and perceived as uncomplicated, what is considered high complexity today may become moderate complexity in the future.

Indeed, major infrastructure projects with a high degree of total complexity have existed for years. In some ways, China's Three Gorges Project, which was completed around 30 years ago, might be regarded as a high-complexity overall project. Additionally, the protracted 30-year discussion over the effect of decision-making in mega infrastructure projects speaks loudly about the need for a full examination of the very complicated decision management and decision-making scheme.

Another representative, albeit more recent, is eastern China's Cross-Bohai Straits Passage Project, which will connect Shandong and Liaoning provinces. This project can also be considered a highly complex mega infrastructure construction project in management and construction technology.

This proposed passage is a complete submerged tunnel with a total length of 123 kilometers that will transport automobiles via train at a speed of 220 kilometers per hour. After completing the Cross-Bohai Straits Passage Project, a new multiplexing system will be created to incorporate both the project and the surrounding society and the natural environment. Thus, the project will unavoidably have a significant, long-lasting, and strong effect on the geological, ecological, social, and economic surroundings within the project region during the building phase and for the duration of the project's operation. Additionally, the project will very certainly raise the probability of severe natural ecological disasters, even though such events have never occurred in the area. This evolutionary risk is difficult to identify and estimate using the typical project demonstration technique. It is incurred by interconnected geology, ocean currents, living animals, the atmosphere, and artificial projects. Thus, the preliminary planning and demonstration of the Cross-Bohai Straits Passage Project must thoroughly identify any significant potential risks to which the project may contribute throughout its long lifespan, including a variety of large-scale evolutionary risks associated with serious natural disasters that may emerge in large numbers. This adds a new and serious dimension to the high complexity of the Cross-Bohai Straits Passage Project's demonstration. Therefore, using the standard project demonstration approach is regarded as unsafe to underestimate this type of risk.

For example, geological conditions in the vicinity of the Cross Bohai Straits Passage Project are highly difficult. For example, the seafloor is rough, with grooves and ridges that descend from west to east and multiple fracture zones on both sides of the Bohai Straits. According to the China

Earthquake Networks Center, on August 22, 2014, regarding the Cross-Bohai Straits Passage Project, the "Tanlu seismic belt extends from the Heilongjiang River in the north to the Yangtze River in the south, runs through the eastern mainland of China from north to east, and spans more than 2400 kilometers." As one of the primary fault zones in northeastern Asia's vast fault system, it has been the site of numerous catastrophic earthquakes throughout history" and is currently active. According to some scientists, the geological state of the seafloor of Bohai Bay is like a broken porcelain plate that retains its original shape, and hence is too delicate to withstand repeated harm and must thus be treated with extreme caution. However, if high-speed trains continue to run through the cross-straits tunnel for an extended period, could this delicate porcelain plate—the Bohai Sea's seafloor—become progressively shattered, resulting in significant geological disasters such as earthquakes? There must be unambiguous responses to the Cross-Straits Passage Project's large-scale evolutionary risk of profound ambiguity. Thus, guided by the new demonstration theory based on the high complexity of mega infrastructure construction, it is important to reach a conclusion that can withstand the test of time and is based on sufficient facts and data produced from a series of scientific discoveries and tests.

In terms of duration, the Cross-Bohai Straits Passage Project is a century-long megaproject. Thus, will the construction process and subsequent operations generate new physical factors and conditions that may accumulate and reinforce over time, resulting in the project settling, cracking, and collapsing due to geological, biological, and current ocean erosion, or will they cause earthquakes? These questions, too, must be subjected to the complex project's quality evolutionary analysis. Through demonstration analysis aided by technologies, such as system simulations of multi-level, multidimensional, multi-scale, and multi-granularity project scenarios, it will be possible to predict phenomena and develop step-by-step laws for managing potential disasters and subsequent phenomena (Mishra et al. 2013; Tolone et al. 2004).

Additionally, the building of the Cross-Bohai Straits Passage Project and its 100-year operation may substantially impact the evolution of the Bohai Sea's marine ecology and cause massive damage to the marine ecosystem and living environment of marine life creatures in the Bohai Bay. These consequences will result from a process of systematic natural disasters transferring, diffusing, and evolving on a broad spatiotemporal scale. Once a disaster occurs, it will significantly impact Bohai Bay, a natural system typified by enclosed seas, a slow hydrodynamic cycle, and a fragile environment. As a result, a modern, multidisciplinary method should be employed for demonstration and study, in which prospective ecological disasters are examined first, and then a dependable contingency plan to avoid or mitigate future disasters is developed.

Additional evolutionary dangers exist for which there is a lack of expertise and an inability to control. If these risks are grossly overestimated during the identification and verification process, the resulting repercussions might be disastrous for the country, society, and natural ecology.

As such, it is critical to fully comprehend that the Cross-Bohai Straits Passage Project is a highly uncertain, high-risk endeavor that is highly integrated in terms of management and construction technology. Additionally, numerous highly interconnected complicated challenges will be addressed for the first time during the project's demonstration in China, or maybe the world at large.

To ensure that the demonstration of the Cross-Bohai Straits Passage Project is of high quality and scientific, it is critical to ensure that the demonstration's primary objective is to identify and forecast large-scale risks and disasters that the project may encounter over its long lifespan, as well as to discover appropriate measures for preventing and mitigating such disasters. As a result, demonstration objectives and a demonstration system must be developed that consider the enormous complexity of this mega-infrastructure installation. To develop the demonstration system, studying the interrelationships between many subsystems, such as society, economy, ecology, and humanities, is necessary.

However, to perform an integrated study of aims, interactions between objectives, conditions, and decision-making norms must be embodied and links formed (Holland 1995; Sheng and Jin 2012).

While certain novel and unique demonstration aims and issues may develop for the first time, these issues and objectives must be well understood and stated, and they should not be simplified or ignored.

Additionally, the Bering Straits Bridge, built by a multinational engineering firm, is a transportation megastructure that will connect Asia and North America. Once completed, it will be another architectural marvel in human history. According to program planning, the bridge will be about 40 kilometers long and include over 220 piers weighing millions of pounds apiece. Their goal is to withstand the huge strain of the Arctic Ocean's deep-water ice, which weighs millions of tons. The bridge is located in an exceedingly hostile environment, more precisely, a chilly climate prone to snowstorms and enveloped in fog. In the winter, temperatures can fall below 45 degrees Celsius, and the straits' surface freezes over with a covering of ice more than 2 meters thick. To construct the bridge in such a harsh environment and overcome obstacles such as colossal icebergs, turbulent seawater, and temperatures as low as 40 °C, the engineering materials, construction machinery equipment, and, most importantly, the perception of relevant construction principles, technical laws, and management must all demonstrate highly integrated complexity and unprecedented risks in the world due to a lack of cognition, information, and knowledge (Stetson and Mumme 2016).

A series of super creation projects envisaged and planned by humanity include mega infrastructure constructions of high complexity, such as the submarine tunnel project under the Bering Straits, which could be up to 105 kilometers in length; the automatic underground freight and railway project traversing the Alps from Rosenheim to Verona, with a total length of single tunnels exceeding 500 kilometers; the tunnel connecting Lyon, France, and Torin, Italy, which is approximately 54 kilometers in length; and an underground railway project traversing the Alps from Rosen.

The management concept applicable to this highly complex mega infrastructure construction may extend beyond what addressed in this book. The complexity of issues must be classified and graded rather than degraded and then effectively coalesced with on-site executive capacity for project management.

For instance, there is the class of fundamental or moderately complicated concerns, referred to as I-level complex issues among the difficult issue classes. Then, based on their complexity, issues are classified as II-level, III-level, and so on, from low to high. Highly complex mega infrastructure construction can be managed at multiple scales depending on the complexity of the issues, with the management goal of gradually degrading complexity from a high to a low level until it coalesces effectively with the on-site executive capacity for project management.

In other words, complexity degradation will remain a critical core principle throughout the management process of extremely complex and integrated mega infrastructure projects, even in the future. However, given the presence of several degrees of complexity, it is advised that the complexity degradation mode be changed from the current once-only mode to a multiple-times, step-by-step method.

THE POSSIBILITIES AND HOPE FOR THE FUTURE

The conventional view of perceiving megaprojects is changing. The new world is shrinking the distance between humans and space, connectivity facilities are improving, and we can see radical transition in human behavior. People are adapting to the new normal, and vice versa. Paul Virilio speaks of the 'end of geography' while others talk of the 'death of distance'. Megaproject infrastructures have gained an alternate anatomy from being a production-consumption architecture to having actual values and impacts in people's lives and its surrounding. Megaprojects are now more focused on telecommunications. It is a tool for internet capitalism and politics of distance. Fascinating innovations are on the way to joining the conglomerate- Megaprojects that will not only help tackle climate change and sustainability issues but can also take humans to Mars or other planets.

Among the futuristics megaproject giants, one is the ITER nuclear fusion project. 35 countries have joined to research and engineer a way for implementing fusion reaction that happens naturally only on Sun. This 65-billion-dollar project is an ambitious endeavor to solve the constantly growing energy demand of the world. Another visionary project is the 100-billion-dollar Forest City in Malaysia. The city connects 4 different islands, and once completed it will have the capacity to house nearly 700, 000 people. What attracts investors to this project is the inclusion of vertical gardens, and technologically advanced methods, integrated to supply every residential need of the modern 21-st century.

In the coming decades, revolutionary megaprojects will change our modes of transportation the most. The China-Pakistan Economic Corridor (CPEC) can be the first key to the revolution. Roughly 2.3 million new job opportunities will arise from the project. In the next 7 years, CPEC is estimated to build 40 new megaprojects, and one of them is Gwadar port. The port will be hosting 200 million tons of cargo annually, which is more than 6 times as much as now. For venturing into deep space, big investment megaprojects have been planned and are ready for execution as well. However, space exploration has always been expensive, and not every nation can afford to invest in such operations.

These are the fine presentiments that indicate the development and change in our society in the next hundred years. It is yet a matter of wonder if these projections will turn us into technology-bound creatures or we will find ourselves amid the steampunk fantasy. The prayers may last with the hope that mankind will find themselves among the greatest glory of time and space.

www.ingramcontent.com/pod-product-compliance
Lightning Source LLC
Chambersburg PA
CBHW070239220526
45465CB00004B/1455